Digital Technologies for Agricultural and Rural Development in the Global South

_____

Digital Technologies for Agricultural and Rural Development in the Global South

---

Edited by

Richard Duncombe

Centre for Development Informatics (CDI), University of Manchester, UK

CABI is a trading name of CAB International

CABI
Nosworthy Way
Wallingford
Oxfordshire OX10 8DE
UK

CABI
745 Atlantic Avenue
8th Floor
Boston, MA 02111
USA

Tel: +44 (0)1491 832111
Fax: +44 (0)1491 833508
E-mail: info@cabi.org
Website: www.cabi.org

Tel: +1 (617)682-9015
E-mail: cabi-nao@cabi.org

A catalogue record for this book is available from the British Library, London, UK.

Library of Congress Cataloging-in-Publication Data

Names: Duncombe, Richard, editor.
Title: Digital technologies for agricultural and rural development in the global south /
    edited by Richard Duncombe.
Description: Boston, MA : CABI, [2018] | Includes bibliographical references and index.
Identifiers: LCCN 2017047041 (print) | LCCN 2017049627 (ebook) |
    ISBN 9781786393340 (pdf) | ISBN 9781786393357 (ePub) | ISBN 9781786393364
    (hardback : alk. paper) | ISBN 9781786394804 (pbk. : alk. paper)
Subjects: LCSH: Agriculture--Data processing. | Information technology.
Classification: LCC S494.5.D3 (ebook) | LCC S494.5.D3 D54 2018 (print) | DDC 338.10285--dc23
LC record available at https://lccn.loc.gov/2017047041

ISBN-13: 9781786393364 (hbk)
ISBN-13: 9781786393340 (pdf)
ISBN-13: 9781786393357 (ePub)

Commissioning editor: David Hemming
Editorial assistant: Alexandra Lainsbury
Production editor: Shankari Wilford

Typeset by AMA DataSet, Preston, UK.
Printed and bound in the UK by CPI Group (UK) Ltd, Croydon, CR0 4YY.

Contents

Figures

Tables and Box

Contributors

Stefano Bocchi holds a PhD in Crop Science from the State University of Milan, where he is a Professor in Agronomy and Cropping Systems. As a Visiting Scientist at the Agronomy Department, University of California, Davis, at IRRI-Philippines, and Wageningen University, he developed research projects on cereals, forage crops, agro-food systems analysis and management. Bocchi is author of more than 180 scientific papers and a Board Member of various scientific societies. He has been involved in several projects for international cooperation in Albania, Brazil, China, Ecuador, Egypt, Kenya, Lebanon, Peru, Philippines, Sierra Leone and Tanzania. He was the scientific curator of the Biodiversity Park in EXPO 2015.
E-mail: stefano.bocchi@unimi.it

Fritz Brugger, whose PhD is in Development Studies, was a founder of Farmforce and was responsible for its overall strategy, conceptual development, project management, piloting and testing from 2010 until market introduction in summer 2013. Subsequently, he has moved to academia and is now a Senior Scientist at ETH Zurich, NADEL Center for Development and Cooperation, Switzerland. Since then, he has had no involvement in Farmforce.
E-mail: fritz.brugger@nadel.ethz.ch

Amanda Caine completed a Masters in Development Policy, Process and Practice at the School of Agriculture, Policy and Development at the University of Reading, where she has recently started a PhD researching into the potential of the internet to empower women in the majority world. She has worked for the last 20 years for a variety of British iNGOs in a finance and strategic planning capacity. She was the Finance Director for several smaller iNGOs (e.g. FARM-Africa) and Head of Planning and Reporting for Oxfam GB. She helped set up GALVmed (Global Alliance for Livestock and Veterinary medicine) and is now a trustee of several iNGOs.
E-mail: a.caine@pgr.reading.ac.uk

Amit Chakravarty is an ICT4D (Information and Communication Technologies for Development) practitioner currently involved in ICT solutions for agriculture with ICRISAT (International Crops Research Institute for the Semi-Arid Tropics) working with smallholder farmers in the drylands tropics across Asia and Africa. Prior to this he led a pan-India ICT4D project funded by UNDP in the domains of rural livelihoods, women empowerment, improving governance and providing citizen services.
E-mail: a.chakravarty@cgiar.org

Chris Clarke is a technology and innovation consultant based at Statistics for Sustainable Development, King's College, London. His current position focuses on how technology can be used to support people and projects in the developing world; investigating ways to enable better data flow and conversion to useful information. This involves designing methods for robust data collection, storage and processing, and finally feeding back through mobile apps, websites, publications, etc. Chris works across multiple countries in Africa and South America on aspects of agriculture and climate.
E-mail: c.clarke@stats4sd.org

Graham Clarkson is a human geographer with 3 years of post-doctoral experience. He completed his PhD, 'Land Use Intensification and Trees on Farms in Malawi', in 2010 at the University of Hull. Graham's primary research interests are based around rural livelihoods and rural change, small-scale agriculture and innovation, communication for development and climate services. He has experience working in East, West and Southern Africa on projects funded by CGIAR centres (IFAD, ICRISAT and CCAFS), USAID, DFID and the UK Economic and Social Research Council, and integrates both qualitative and quantitative research methods with expertise in participatory approaches for research and practical application. Graham's current research involves collaboration with colleagues from academia, government and non-government organizations in Uganda, Kenya, Tanzania, Ghana, Lesotho, Mali, Sudan, Burkina Faso and Malawi.
E-mail: g.clarkson@reading.ac.uk

Charlotte Day is Nutrition Project Manager at CAB International, leading on the mNutrition Initiative and supporting business development in nutrition-sensitive agriculture. Previously, Charlotte completed an MSc in Food Security at the University of Warwick, including research on crop stress and water use efficiency at the Eden Project in Cornwall. Following this Charlotte worked in Kenya for an NGO, delivering training to support nutrition-sensitive agriculture interventions with smallholder farming families. Charlotte holds a BASIS Certificate in Crop Protection and is a FACTS Qualified Fertilizer Adviser.
E-mail: C.Day@cabi.org

Andrew Dearden is Professor of Interactive Systems Design at Sheffield Hallam University and has published widely on how effective design practices, specialist design skills and good design ideas can be shared to allow people who are not specialist designers to devise workable and appropriate systems. His work is particularly concerned with enabling people and groups who may have limited resources, or limited experience with technology, to shape systems for themselves. He has worked with private, public and third-sector organizations in the UK, India and Africa. In 2009, his team's work with a cooperative of marginalized Indian farmers received an Award from the Manthan Foundation (South Asia) for its contribution to e-Enterprise and Livelihood.
E-mail: a.m.dearden@shu.ac.uk

Peter Dorward has experience in a wide range of areas of research, practice and innovation for smallholder agriculture including climate services, smallholder adaptation and resilience to climate variability and change, understanding decision making in smallholder households and implications for innovation and support, participatory approaches for facilitating farmer innovation in smallholder farming systems, and many more. Peter has developed a novel climate services approach, Participatory Integrated Climate Services for Agriculture (PICSA), which is becoming widely used in developing countries. International organizations supporting and using the approach include the World Food Programme (WFP), CCAFS, and the International Fund for Agricultural Development (IFAD), working with government departments and NGOs at country level.
E-mail: p.t.dorward@reading.ac.uk

Richard Duncombe completed a Masters in Economics and Innovation at the Science Policy Research Unit (University of Sussex) where he also worked as a Research Officer. He took his PhD at the University of Manchester within the Global Development Institute (GDI). Previously, Richard lived in Africa working in the field of ICT4D (information and communication technologies for

development) in Tanzania and Botswana for most of the 1990s before moving to the University of Manchester as a Lecturer in 2000. Richard works at the Centre for Development Informatics (CDI), University of Manchester, the largest academic group dedicated to ICT4D work in the UK.
E-mail: richard.duncombe@manchester.ac.uk

Gordon A. Gow is Associate Professor of Communication and Director of the Graduate Program in Communications and Technology (MACT) in the Faculty of Extension at the University of Alberta. Dr Gow's research interests revolve around the social impact of information and communication technologies (ICTs) in the areas of community engagement, public health, agriculture, and international development. He has been visiting Sri Lanka since 2006, collaborating on various research projects with LIRNEasia and Wayamba University of Sri Lanka.
E-mail: ggow@ualberta.ca

Stan Karanasios is currently a Vice Chancellor's Senior Research Fellow in the College of Business at RMIT University in Melbourne, Australia. After completing his PhD at Victoria University in Melbourne he worked at the University of Leeds for six years, where he currently holds a visiting role. A key aspect of his research is how technology transforms, disrupts and compliments human activity, particularly in the context of ICT4D. His research works spans Europe, Asia, South America, Africa and Australia. He has published in leading information systems journals, including, among others, *MIS Quarterly, Information Systems Journal, European Journal of Information Systems* and *Journal of the American Society for Information Science and Technology.* In addition to his research work he has undertaken a range of consulting work for industry and international bodies, such as the United Nations.
E-mail: stan.karanasios@rmit.edu.au

Linus Kendall is currently a PhD researcher at Sheffield Hallam University where he is working on ICT4D research focused on design of agricultural knowledge management systems for smallholder and subsistence farmers. He has previously worked with sustainable development in various sectors, from for-profit to social enterprises to non-profit. Most recently he has lived and worked in India in both urban and rural development, with a specific focus on climate change and sustainable development.
E-mail: b5035879@hera.shu.ac.uk

Luis Emilio Lastra-Gil is pursuing a PhD in Information Systems and Innovation within the Department of Management at the London School of Economics (LSE). He holds an MSc in Management from Tecnológico de Monterrey (ITESM) in Mexico, an MBA from the University of Bath, an MSc in Economic History (Research) from LSE, and a Diploma in English History and Literature from the University of Oxford. Emilio has more than 15 years of experience in telecommunications in roles that include quality and business strategy, project management, bids and proposals, and more recently business modelling and business analysis for Bell Labs Consulting.
E-mail: L.E.Lastra-gil@lse.ac.uk

Alberto Lubatti has a Masters degree in Agricultural Science at the University of Milan, for which he presented a thesis entitled 'Farming system and ICT adoption. The case of Sierra Leone'.
E-mail: alberto.lubatti@gmail.com

Christopher A. Moturi is currently the Director at the ICT Centre of the University of Nairobi. Prior to that, he was Deputy Director, School of Computing and Informatics, where he has worked since 1986. He is an experienced university manager and his roles include management of the ISO9001 Quality Management System, Lead Internal Quality Auditor, Head of Department, Course Coordinator, Examinations Officer; member of the University Management Board, Senate, Dean's Committee, College Management Board and College Academic Board; and delivering teaching, curriculum development, research and consultancy.
E-mail: moturi@uonbi.ac.ke

Mukund D. Patil is an ICT4D (Information and Communication Technologies for Development) scientist currently involved in ICT solutions for agriculture with ICRISAT (International Crops Research Institute for the Semi-Arid Tropics) working with smallholder farmers in the drylands tropics across Asia and Africa.
E-mail: M.Patil@cgiar.org

Andrea Porro completed a Master of Science in crop production at Department of Agricultural and Environmental Sciences – Production, Landscape, Agroenergy (DISAA) of the University of Milan, and received a PhD in agricultural sciences from the same university. His main interest as a researcher and practitioner has been the contribution of GIS and agrotechniques to agricultural development, and natural resources management in the domain of agroecology. He is currently serving as country representative for iMMAP (www.immap.org), a not-for-profit organization involved in supporting UN agencies in emergency operations with Information Management Services in the field of food security, health and protection for people in need.
E-mail: aporro@immap.org

Simone Sala completed a Master of Science in Information and Communication Technologies at the Computer Science Department of the University of Milan, and received a PhD in agricultural sciences from the same university. He hold research assignments at Columbia University, University of Milan, and University of Lugano. His main interest as a researcher and practitioner has been the contribution of ICTs to agricultural development and natural resources management. He is currently serving at the UN Food and Agriculture Organization (FAO) as an adviser on ICT4D, with a specific focus on family farming.
E-mail: simone@simonesala.it

Worlali Senyo is a Senior Consultant at Farmerline and is currently based in Accra, Ghana. He manages Farmerline's client and partner relationships by organizing meetings and presentations of Farmerline's products and services. Worlali proactively identifies fundraising and business opportunities for Farmerline. Before joining the Farmerline team, Worlali worked at the Technical Centre for Agricultural and Rural Cooperation ACP-EU (CTA). He has over 10 years' experience in the field of Agriculture and ICTs relating to mobile technologies for agriculture and rural development with a focus on project planning, design, implementation, monitoring and evaluation around thematic areas such as: food security, knowledge management, agribusiness, community informatics and innovation, and agriculture policies. He holds a Masters degree in ICTs for Development from the University of Manchester, UK and a Bachelor's degree in agricultural science majoring in economics from the University of Ghana.
E-mail: worlali@farmerline.org

Mira Slavova has extensive experience in technology innovation within the African context (Liberia, Ghana, South Africa). She holds a PhD from the University of Cambridge and has completed post-doctoral research at Leeds University Business School and the International Food Policy Research Institute (Washington, DC). Mira is currently an associated researcher at the Gordon Institute of Business Science (University of Pretoria) and a guest lecturer at ESSEC Business School (Paris). Her academic interests are focused on ICTs as institutional carriers, 'third spaces' for knowledge interactions, design-led innovation and platform design.
E-mail: slavovam@gibs.co.za

V.V. Sumanthkumar is an ICT4D (Information and Communication Technologies for Development) scientist currently involved in ICT solutions for agriculture with ICRISAT (International Crops Research Institute for the Semi-Arid Tropics, Hyderabad, India) working with smallholder farmers in the drylands tropics across Asia and Africa.
E-mail: sumanth@naarm.org.in

Luka M. Wanjohi completed his Master of Science degree in Information Technology Management at the University of Nairobi, Kenya. Luka works at the International Potato Center, Nairobi as a Senior Regional Knowledge Management Associate. He is involved in development and management of information systems for sweet potato research work in more than 10 countries in sub-Saharan Africa.
E-mail: l.wanjohi@cgiar.org

Abbreviations

ADRA	Adventist Development Relief Agency
B2B	Business to Business
BMJ	British Medical Journal
CABI	CAB International
CCAFS	Climate Change, Agriculture and Food Security
CGIAR	Consultative Group for International Agricultural Research
CKW	Community Knowledge Worker
CRM	Customer Relationship Management
DfID	Department for International Development (UK)
DOEA	Sri Lanka Department of Export Agriculture
ERP	Enterprise Resource Planning
FAO	Food and Agriculture Organization
GAIN	Global Alliance for Improved Nutrition
GCP	Global Content Partners
GPS	Global Positioning System
GSMA	Global Standards Mobile Association
ICRISAT	International Crops Research Institute for the Semi-Arid Tropics
ICT4D	Information and Communication Technology for Development
IFPRI	International Food Policy Research Institute
IKSL	IFCO Kisan Sanchar Ltd
ILRI	International Livestock Research Institute
INGO	International Non-Government Organization
IVR	Integrated Voice Response
JETI	Joint Education and Training Initiative
KGS	Krishi Gyan Sagar
KPI	Key Performance Indicator
LCP	Local Content Partners
MFI	Micro-Finance Institution
MNO	Mobile Network Operator
MoSCoW	Must have, Should have, Could have and Won't have
MRL	Minimal Residual Limits
NGO	Non-Government Organization

PICSA	Participatory Integrated Climate Services for Agriculture
QA	Quality Assurance
QC	Quality Control
RAM	Random Access Memory
RUTAM	Rural Technology Acceptance Model
SaaS	Software as a Service
SLARI	Sierra Leone Agricultural Research Institute
SMS	Short Messaging Service
SSHRC	Social Sciences and Humanities Research Council of Canada
SUN	Scaling Up Nutrition
TRAI	Telecom Regulatory Authority of India
UNIMAK	University of Makeni
USAID	United States Agency for International Development
VAS	Value Added Service
VSS	Voluntary Sustainability Standards
WHO	World Health Organization

Introduction and Overview

Richard Duncombe

Centre for Development Informatics, University of Manchester, UK

The population of the Global South is growing rapidly with the largest countries by population, such as China and India, and the poorest countries by GDP per capita, such as those in sub-Saharan Africa, showing the highest predicted population growth rates according to the World Bank in 2016. As populations grow so do requirements for food and nutrition. In this respect, improvements in agricultural productivity and sustainability have always been essential conditions for development to take place. Agricultural productivity (measured as agricultural output per capita) has risen steadily in the developing world over the last five decades (Benin *et al.*, 2011). For many countries this has been a great success story, as improved agriculture has the greatest impact on the reduction in rural poverty. However, some regions of the Global South have seen greater success than others.

A weakness in many poorer developing countries is that agricultural growth has been driven, to a large extent, by expanding land use – including de-forestation. Productivity has increased at a much slower rate than the expansion of land given over to agricultural production. In order to curtail the use of land, which is a finite and valuable resource, increased food production must rely on increased yields through improved seeds, fertilizers and irrigation. Another drawback of insufficient domestic production is a growing reliance on imported food. The poorest region – sub-Saharan Africa – spends about US$30 billion to US$50 billion a year to import food. As a result, the continent lacks funds to invest in infrastructure, social and economic amenities. If domestic production does not increase dramatically, Africa is likely to spend around US$150 billion on food imports by 2030 (Fuglie and Rada, 2013).

Low agricultural productivity in the Global South has many causes that relate to lack of knowledge of up-to-date technologies and practices: low use of improved seed, use of inappropriate fertilizer, inadequate irrigation and lack of incentives for farmers in the absence of remunerative markets. In many countries, these sector-specific problems are compounded by overarching issues, including political instability and violent conflict, weak institutions of governance and ineffective policies or rural people's poor health. Climate change is also likely to exacerbate matters further. For instance, the yields from rain-fed agriculture are set to decline in some countries (Fuglie, 2012). Therefore, farmers may need adaptation strategies like water harvesting, cultivating drought-resistant crops and ecological restoration.

In recent years, to face these challenges, there has been growing activity around use of digital technology for agricultural and rural development in the Global South, to address the knowledge gaps discussed previously, and to establish the building blocks for new rural services. The introduction of mobile phones and the internet into rural areas may be one means to improve productivity and counteract scale-economies for individual producers, by strengthening existing agricultural knowledge systems and stimulating forms of intervention incorporating new innovative services (Qiang *et al*, 2011; Deichmann *et al.*, 2016). Mobile phone-based services have proliferated in recent years and provide new ways to access price and market information, and coordinate input/output resources including transport and logistics, finance and production techniques. Personal use of the mobile phone has also enabled rural producers to interact directly with end-user markets, traders, suppliers, extension services and with each other (Duncombe, 2016).

The aim of this book is to share research and practice on current trends in digital technology for agricultural and rural development in the Global South: specifically to bring together the perspectives of academic researchers from diverse disciplines with those of practitioners with experience of implementing mobile applications and agricultural information systems in differing country contexts. Growth of research and evaluation in this field has been slower than the pace of change for practitioners, and much remains to be done, particularly in successfully combining the opportunities afforded by digital development with the socio-economic realities of agricultural transformation in developing countries. The book endeavours to close this gap and is organized into three parts.

Part 1 focuses on the **creation and sharing of knowledge**. The focus on agricultural knowledge raises important questions, which are addressed by the contributors. Principally, what are the ways in which agricultural knowledge can be turned into effective practice, and what role can digital technologies play in this process?

In Chapter 1, **Amanda Caine**, **Chris Clarke**, **Graham Clarkson** and **Peter Doward** from the **University of Reading** investigate the Participatory Integrated Climate Services for Agriculture (PICSA) that supports decision making aims to build resilience amongst smallholder farmers in Africa. Analysis of 'proof-of-concept' mobile-phone applications shows how the PICSA approach is used to provide support to existing knowledge-based climate and agricultural decision making. The chapter sets out the PICSA approach before summarising the findings of a review that was undertaken to highlight the lessons learned from similar mobile-agriculture (mAgri) initiatives. Two mobile application case studies, which focus on historical climate information and participatory budgeting for smallholders in Northern Ghana, are examined and discussed. These initiatives are found to have the potential to scale the provision of agriculture-related information and services to a large number of smallholder farmers at a relatively low cost.

Chapter 2 focuses on the collection of monitoring data, and integration of data into agricultural program management activities. **Luka Wanjohi** from the **International Potato Center** and **Christopher A. Moturi** from the **University of Nairobi** share experiences of the use of smartphones to collect monitoring data for sweet potato vine multiplication and dissemination activities in sub-Saharan Africa. An Open Data Kit (ODK) technology is used to collect geo-tagged data, perform error validation, on-site data quality control, capture photos and much more, using Android mobile devices. The chapter outlines how monitoring and evaluation (M&E) personnel from countries across sub-Saharan Africa received training on how to develop the electronic forms and use them for data collection, which, over a period of two years, captured new registration data for 326 vine multipliers in Ethiopia, Ghana, Kenya, Malawi, Mozambique, Nigeria and Uganda. The case study demonstrates enhanced efficiency and how the ODK system has boosted efforts to standardize the data collection across regions. Constraints related to user take-up and comparative costs are also addressed.

In Chapter 3, **Amit Chakravarty, V.V. Sumanthkumar** and **Mukund D. Patil** , from the **International Crops Research Institute for the Semi-Arid Tropics (ICRISAT)**, Hyderabad, India, present a case study from India of a knowledge database which assembles a wide range of agricultural data relating to land holding, soil analysis, cropping patterns, past yield and fertilizer use. The case demonstrates how mobile phones are used to relay information to farmers on weather, market prices, crop management practices, disease management, and many other aspects. Particular attention is paid to investigating how a multiplicity of agencies, both government and private, can make use of such data. Because agencies and farmers themselves require tailored information packages, the 'broadcasting' of information is of limited value; rather 'narrowcasting' targeted information serves the farmers' needs better and it is more likely that the farmers will act upon the advice.

The chapter assesses the use of tablets made available to farmer facilitators, for uploading/updating farmer data as well as downloading and sharing information with the farmers, and relaying information to experts in other locations.

In Chapter 4, **Charlotte Day** from **CAB International** provides an early assessment of the GSMA's mNutrition Initiative, funded by DfID, which is working to bridge the business interests of mobile operators with this development aim, by working with mobile operators and other partners in 12 countries in Africa (Ghana, Nigeria, Malawi, Tanzania, Mozambique, Uganda, Kenya and Zambia) and South Asia (Sri Lanka, Bangladesh, Pakistan and Myanmar) to provide commercially sustainable agriculture and health value-added services to the rural poor. The chapter focuses on the content-related experiences and lessons of the mAgri component to mNutrition, as reported by content partners of the initiative. A particular focus of this initiative was to test the feasibility of mobile communications for behaviour change in agricultural practices, especially related to the impact this has on household nutrition.

Part Two examines **knowledge intermediaries**, the organizations or individuals who stand between the farmers themselves and the markets and institutions with whom they interact (e.g., suppliers of inputs and finance, buyers of produce and providers of knowledge and assistance). The contributions are centrally concerned with how mobile technologies effect the agency and capabilities of end users to act on content provided through digital channels, and how producers are able to change their behaviour on the basis of data and information that can be accessed via digital technologies – in essence, understanding the mechanisms that can turn information into usable knowledge.

In Chapter 5 **Gordon A. Gow** from the **University of Alberta** puts forward 'technology stewardship' as an approach for training and supporting individuals and teams who engage agricultural communities to encourage and support innovative practices with low-cost, widely available digital technologies. The chapter draws upon theoretical ideas of 'communities of practice' to analyse a stewardship initiative that has been implemented in Sri Lanka, presenting preliminary results of an ongoing action research project involving pilot studies of technology stewardship with partner organizations. The chapter outlines plans for a Joint Education and Training Initiative for technology stewardship that includes community engagement, rapid prototyping, evaluation and impact assessment, and was launched in September 2016.

In Chapter 6, **Fritz Brugger** from **NADEL Center for Development and Cooperation, Switzerland** focuses on contract farming as an approach to support smallholders who are aiming to produce high-value crops for domestic formal markets and for export. The chapter argues that contract farming can be an effective institutional mediator for smallholders, which helps to raise their productivity and orient their production toward more remunerative commodities and markets. A case study of contract farming which uses a mobile phone-based Software-as-a-Service (SaaS) business management system is evaluated. Mobile-phone technology is observed to be a key driver of change that is stimulated by contract farming, which enables a reduction in overall transaction costs for the farmers. The chapter sets out detailed guidance on best practice for implementing such initiatives and highlights a number of pitfalls relating to how pilot projects are evaluated, appropriate business models, and efforts to scale and sustain interventions.

In Chapter 7, **Simone Sala**, **Andrea Porro**, **Alberto Lubatti** and **Stefano Bocchi**, from the **DISAA, Università degli Studi di Milano** investigate the relative importance of the mobile phone, the internet and social media as mediating tools for rice farmers in Sierra Leone, given that rice forms the backbone of the national food security system. Researchers from the University of Milan and the local University of Makeni carried out primary research, collecting data from farmers and other respondents in 2014, with a focus on different ways in which farmers accessed and exchanged information during different stages of the growing season, and across different media channels. The study highlights very dynamic behaviour of farmers, and identifies a high diffusion of mobile phones and relatively high access to the internet among rice farmers in the area of study. Variables such as income were correlated with factors such as availability and type of use of ICT products and services among rice farmers, but in common with other studies, face-to-face interaction was not found to have been diminished significantly due to the advent of ICT-based tools.

In Chapter 8, **Luis Emilo Lastra-Gil** from the **London School of Economics** uses a lens of institutional economics to present a study of the role of ICTs in distribution channels for agricultural produce in Mexico. The focus on distribution emphasizes the importance of market knowledge, entrepreneurial skill, and expertise in negotiation and trading, whilst also requiring that capital be invested to build knowledge and appropriate networks directly through professional managers with a deep knowledge of market intermediaries, or alongside other farmers in a cooperative or marketplace, or with customers via a collaborative supply chain network. ICTs are observed to be central to this process, establishing new links between farmers, markets and end consumers, and optimizing distribution channels that potentially reduce transaction costs. Thus, ICTs are primarily used by farming communities as a management tool for intermediation and cooperation by building relationships, and through acquisition of knowledge.

Part 3 provides different perspectives on **how digital technologies can facilitate change in agricultural systems**. A central theme relates to 'disruption' versus 'sustainability' of existing agricultural structures and processes. A key question examined is what kinds of institutional impacts are produced by content (knowledge) diffusion initiatives, and what are the challenges of successfully bridging business and development interests in the agricultural sector through use of digital technologies?

In Chapter 9, **Linus Kendall** and **Andrew Dearden** from **Sheffield Hallam University** explore alternative theories of change for mobile technologies for agriculture and rural development. They provide a critique of popular approaches that employ 'theory of change' within the field of international development, where it has been used by both international donors and civil society. It is suggested that explicitly acknowledging, and critically examining, a theory of change helps to illuminate the assumptions, conditions and processes by which an intervention seeks change, which is valuable in all phases of a project, from design to evaluation. The chapter looks at two theories of change critically and highlights some of the challenges they have faced in practice. In response to these challenges, an alternative approach to developing a theory of change is suggested, which draws upon human development theories. A case study of an ongoing research project exemplifies such an alternative theory, and the implications are set out for the design of ICTs in agricultural settings.

In Chapter 10, **Richard Duncombe** from the **University of Manchester** suggests that the role of mobile-phone-led services for agricultural and rural development needs to be viewed in a enabling context that highlights both systemic and organizational change. The chapter tracks the evolution of a mobile-phone-led service from East Africa. The case study demonstrates that the process of changing agricultural systems is complex, and requires considerable organizational effort as well as the necessary seedcorn financing. New digital intermediaries are found to be substituting both for the top-down role of the state (by providing an alternative to the traditional government extension services), and the bottom-up role of farmer collective action (traditionally organized through farmers' cooperatives). Thus, local enablers of innovation processes are found to be crucial in the development of successful mobile applications and the evidence suggests that achievement of scale through collective action is an effective and productive way to transform pre-existing farming systems.

In Chapter 11, **Stan Karanasios** and **Mira Slavova** from the **Royal Melbourne Institute of Technology** and the **Gordon Institute of Business Science** examine information practices and ICT use in rural Ghana. The study shows how mobile phones introduced among smallholder farmers create 'hybrid' information practices that are consistent with both existing cultural-historical norms around farming and ICT use (a smallholder logic) and with policy imperatives aimed at re-casting farming 'as a business' and promoting value-chain integration through ICT (a value chain logic). Additionally, the study observes how actors in rural agriculture and agriculture partners are able to leverage 'hybrid' practices in advancing their value chain agenda. In designing interventions, these actors put forward 'hybrid' approaches, combining technologies, which align with the rural oral tradition with ones that correspond to contemporary business norms (e.g., mobile phones).

Chapter 12 presents an overview by **Worlali Senyo** of **Farmerline** – a for-profit agtech company in Ghana that is facilitating change through the application of ICT. The enterprise was founded after Mobile Web Ghana's Apps Competition in 2011, organized by the World Wide Web Foundation, where the two co-founders, Alloysius Attah and Emmanuel Owusu Addai paired to develop a solution to send farming tips to farmers via text messages. Their application was selected winner of the competition and won a prize of US$600, which became the start-up capital for the company. This case study is emblematic of a plethora of new digital start-ups that are creating new opportunities to deploy and exploit ICTs in the agricultural sectors of developing countries.

In the final Chapter 13, **Richard Duncombe** summarizes the lessons for best practice based on the cases reported in the chapters, puts forward a themed collection of possible future research questions, and provides a list of sources of further information.

References

Benin, S., Nin Pratt, A., Wood, S. and Guo, Z. (2011) Trends and spatial patterns in agricultural productivity in Africa, 1961–2010. *ReSAKSS Annual Trends and Outlook Report 2011,* International Food Policy Research Institute (IFPRI), Washington DC. Available at: http://www.resakss.org/sites/default/files/pdfs/trends-and-spatial-patterns-in-agricultural-produc-53115.pdf (accessed 15 December 2015).

Deichmann, U., Goyal, A. and Mishra, D. (2016) Will digital technologies transform agriculture in developing countries? *Agricultural Economics* 47 (S1): 21–33.

Duncombe, R.A. (2016) Mobile phones, agricultural and rural development: a literature review and future research directions. *European Journal of Development Research*, 28, 213–235.

Fuglie, K. (2012) Productivity growth and technology capital in the global agricultural economy. In: Fuglie, K., Wang, S.L. and Ball, V.E. (eds) *Productivity Growth in Agriculture: An International Perspective.* CAB International, Wallingford, UK.

Fuglie, K.O. and Rada, N.E. (2013) Resources, policies and agricultural productivity in sub-Saharan Africa. *Economic Research Report No 145*, United States Department of Agriculture, Economic Research Service, Washington, DC. Available at: https://www.ers.usda.gov/publications/pub-details/?pubid=45047 (accessed September 2017).

Qiang, C.Z., Kuek, S.C., Dymond, A. and Esselaar, S. (2011) Mobile applications for agriculture and rural development. ICT Sector Unit, The World Bank, Washington DC. Available at: http://siteresources.worldbank.org/INFORMATIONANDCOMMUNICATIONANDTECHNOLOGIES/Resources/MobileApplications_for_ARD.pdf (accessed 17 January 2014).

1 Mobile Phone Applications for Weather and Climate Information for Smallholder Farmer Decision Making

Amanda Caine,[1] Chris Clarke,[2] Graham Clarkson[1] and Peter Dorward[1]*

[1]School of Agriculture, Policy and Development, University of Reading, UK; [2]Statistics for Sustainable Development, Reading, UK

1.1 Introduction

Smallholder farmers are particularly vulnerable to climate variability and change. Providing smallholder farmers with climate information can enable them to make better farming decisions which can in turn lead to increased food security. The Participatory Integrated Climate Services for Agriculture (PICSA) approach (Dorwood *et al.*, 2015) seeks to support decision-making and build resilience amongst smallholder farmers in Africa by providing climate information and decision-making tools. Whilst it has been successful to date and reached tens of thousands of farmers, greater use of mobile phones and apps to support PICSA may have the potential to enhance certain aspects. Mobile phones are being used increasingly to provide smallholder farmers with agricultural information and advisory services, with a wide variety of mAgri initiatives being established in the developing world over the last few years. These initiatives offer the potential of providing agriculture-related information and services to a large number of smallholder farmers at a relatively low cost. This chapter considers how mobile phone applications may be used to provide weather and climate information to smallholder farmers. In particular, it discusses the development and testing of two proof-of-concept mobile phone applications that use elements of the PICSA approach to provide climate information and decision-making tools. The chapter starts with a brief explanation of the PICSA approach before summarising the findings of a review that was undertaken to highlight the lessons learned from existing mAgri initiatives and inform the development of the mobile applications. The two mobile applications, which focus on historical climate information and participatory budgeting, are then described and the initial observations and findings from a proof-of-concept project in Northern Ghana are examined and discussed. Future developments of the two mobile applications are considered and further research questions posited.

1.2 Participatory Integrated Climate Services for Agriculture (PICSA)

PICSA is an approach that seeks to build resilience at the farm level by supporting

---

* Corresponding author e-mail: p.t.dorward@reading.ac.uk

decision-making through the integration of information on location-specific climate, crops, livestock and livelihoods (Dorward *et al.*, 2015). It emphasizes practical, hands-on methods that can easily be used and understood by farmers. The approach involves agricultural extension staff or community volunteers working with established groups of farmers ahead of the agricultural season to jointly analyse historical climate information and use participatory tools to develop and choose crop, livestock and livelihood options best suited to individual farmers' circumstances and the local climate. Closer to, and within the season, farmers may make adjustments to these plans with the aid of forecasts. Currently relatively little use is made of mobile phones in PICSA.

1.2.1 History of PICSA

Work on developing the PICSA approach started in 2011 at a small scale in Zimbabwe, and further research and development through work in Kenya and Tanzania brought it to a stage that, in 2015, it was scaled out to more than 10,000 farmers across three countries in sub-Saharan Africa (Ghana, Malawi and Tanzania). Building on this, work has continued and at the time of writing PICSA training has been carried out on a pilot basis in Colombia, Senegal, Burkina Faso and Mali, and in Rwanda the approach will reach all 30 districts by the end of 2018.

Good partnerships already exist with NGOs including Oxfam, ADRA and CARE as well as with the required government services. The team are regularly asked to support work in new locations, and requests have recently been made for PICSA to be implemented in Lesotho, Zambia, Niger and other countries including in the Caribbean, Latin America and Asia. Ahead of operating in new countries, groundwork is necessary on meteorological data, identifying agricultural options and training staff. The team are continually improving the approach and this includes learning from feedback and innovation, improving components such as analysis of crop probabilities, working on how to make better use of mobile phones, incorporation of satellite data to provide historical climate information where rain gauge records are unavailable, and

developing additional training materials such as videos.

1.2.2 An explanation of the approach

Starting with the farmer at the centre, PICSA has three key components that are developed to encourage an integrated approach to extension. The three components are: climate information; crop, livestock and livelihood options; and participatory decision-making tools.

Climate information

This is made up of locally specific historical climate information and locally specific forecasts on both seasonal and short-term timescales. This involves considerable work and capacity development with national meteorological services. The climate and weather information is then 'packaged' and communicated using simple graphs that are useful and useable for extension staff, community volunteers and farmers. Farmers are able to use this information to examine and explore whether and how their climate is changing and, importantly, the variability in the weather conditions they experience, as well as to obtain better understanding of how different forecasts can be interpreted and may help in decision making.

Options

This component involves several steps. Preparation in advance of training of extension staff or volunteers who will work with farmers can help to identify potential options that may be available in specific locations and that help to address climate challenges. In the process of training trainers the extension workers or community volunteers are challenged to consider the different options that may be suitable for farmers in their location. Options may include new enterprizes and changes to management of existing ones (e.g., new crops, new livestock or other livelihood activities such as starting a new business, and changes to management practices through different crop varieties, planting dates, soil and water management practices, and use of veterinary care). During the roll out, farmers in groups are facilitated by extension staff to discuss and explore their options in a structured exercise.

Participatory decision-making tools

PICSA involves a set of participatory tools to enable farmers to analyse and use the locally specific climate information and then consider their options in the context of their local climate. After considering the information and options, farmers are able to use tools like participatory budgets to plan and make decisions about their farming and livelihoods for the coming and future seasons.

1.2.3 Key principles of PICSA

The key principles behind the PICSA approach are that the 'farmer is the decision maker' and to provide 'options by context'. As stated above, the PICSA approach keeps the farmer at the centre. This includes putting emphasis on farmers making their own decisions and contrasts with some extension approaches that often place more emphasis on providing 'advisories' or telling farmers what to do. PICSA looks to provide evidence for decision making and a set of tools that can help farmers to interpret this and then to make their own plans and decisions.

'Options by context' is the understanding that all farmers are individuals and have different 'contexts': different educations, access to resources, attitudes to risk and goals *inter alia*. Options that interest and/or will be of use to one farmer may be very different to another even if they are neighbours.

1.2.4 PICSA and mobile applications

PICSA has proven to be successful both at the pilot scale and is successfully going to large scale in several countries, as noted earlier. The participatory nature of interactions between facilitators and farmers is an important part of PICSA, and PICSA makes relatively little use of mobile phones and none of the mobile applications at present. There may be potential to make better use of, and to take advantage of, the increasing availability of low-cost smartphones and tablets

in supporting PICSA implementation. These devices provide an opportunity to reach smallholder farmers on a large-scale with locally specific information and tools that are useful for their decision making. However, their ability to reach farmers on a large scale does not make mobile applications a panacea. As a first step in considering the potential for mobile applications to support delivery of PICSA, a review of a range of initiatives and projects that have been implemented using mobile applications for agricultural decision making was conducted.

1.3 Review of mAgri Initiatives

As explained above, the purpose of the review (Caine *et al.*, 2015) was to investigate the use of mobile applications to provide smallholder farmers with weather and climate information as well as any weather-related learning, advisory and extension services in respect of crop production. The review aimed to ascertain the types of weather-related information that were being provided to identify lessons to be learned from these initiatives and to consider their key factors for success. These lessons learned and factors for success would then be incorporated into a proof-of-concept project to develop mobile applications to provide farmers in the north of Ghana with weather- and climate-related information.

Although the initial intention was to look at the use of mobile applications within sub-Saharan Africa, the geographical scope of the review was broadened to include initiatives in India as a consequence of the greater penetration of mAgri initiatives and the increased body of research on these initiatives. The review also included the use of mobile applications on phones, tablets and phablets.

The review was based on the literature, 15 key informant interviews and a case study analysis of 15 initiatives that have used mobile applications with smallholder farmers. The key informants worked for a range of organizations that are involved in the mAgri sector in Africa and India, such as non-governmental organizations (NGOs), mobile network operators, multilateral agencies, industry associations and private companies and included USAID, CAB International, Oxfam GB, Bill & Melinda Gates

Foundation and Vodafone. It was not intended that the informants would be representative of the whole of the sector, but they were selected on the basis of their knowledge of certain aspects of it. The initiatives used for the case studies were selected because they displayed particularly interesting, unusual or successful features, had a focus on weather-related information or showcased a particular business model or partnership. The initiatives are shown in Table 1.1 below. The case study analysis was based on information found in grey literature and from discussions with key informants.

1.3.1 Establishing user needs

The literature review, key informant interviews and the case study analysis all highlighted the importance of ascertaining farmers' specific information requirements and understanding how information will be used by farmers within

their local context whilst or before developing the application. This entails engaging with farmers and local communities to ensure that the content is relevant to them, relates to their own knowledge base and the information can be easily accessed, assimilated and applied by them. The content and the design of the mAgri products must therefore take full cognizance of the educational attainment, gender, age and the informational and technological skills of their target audience as well as the local circumstances in which they are being used (Masuki *et al.*, 2010; Kameswari *et al.*, 2011). All too often the focus is on what the technology can deliver and the 'perceived' needs of the farmers by outsiders, rather than developing an in-depth understanding (Hellström, 2010; Glendenning and Ficarelli, 2012). The review therefore reveals how, ideally, mAgri applications should be developed with farmer involvement. Indeed some of the more successful mAgri initiatives, for example, IKSL, have reaped the benefits from their human-centred design approach, with

Table 1.1. List of case studies.

mAgri Initiative	Country	Main information and services provided
Digital Green	Ethiopia, India	Rural livelihoods information
Esoko	Ghana	Market prices, weather forecasts, agricultural tips
Farmerline	Ghana	Weather forecasts, agricultural news, pest alerts, crop prices
Reuters Market Light	India	Extensive crop information, market prices, detailed weather information
IKSL	India	Crop information, horticulture, animal husbandry, news alerts, weather forecasts, entomology
Green Phablet	India	Weather information, forecasts, pest information, crop and agricultural input prices, expert talks, learning packages
MKisan	India	Crop information, market prices, animal health, weather forecasts
Kilimo Salama (now ACRE)	Kenya, Tanzania	Agronomy and agricultural meteorology, weather forecasts
Airtel Kilimo	Kenya	Weather forecasts, crop information, market information
Senekela	Mali	Agronomy, market information
Tigo Kilimo	Tanzania	Weather forecasts, agronomy tips, market price information
Beep4weather	Tanzania	Weather forecasts, agricultural meteorological information
Sesame Marketing Project	Tanzania	Information about sesame production and marketing
Community Knowledge Worker	Uganda	Weather information, market prices, crop and livestock management
Agri-Fin Mobile	Zimbabwe, Uganda	Crop information, weather information, financial services

increased and more sustained farmer engagement with their mAgri products and services over the long term.

1.3.2 Content

The provision of localized content that is accurate, credible and reliable is a significant challenge for mAgri initiatives. Some initiatives have been able to develop partnerships with local organizations who can produce such high-quality content whereas others, such as Esoko, have developed their own content. The provision of localized content and the maintenance of its quality control is expensive, and this partly explains the paucity of sustainable mAgri business models. The integration of different types of complementary information into 'bundles' is a way in which the service can become financially sustainable and can also make the information become more actionable by the farmer. In particular, it provides a way in which more profitable services, such as micro-insurance or financial services, can subsidize less profitable services, such as weather or crop information.

1.3.3 Timeliness and context

For the information to be relevant and actionable by farmers, it must also be timely. Some initiatives, for example Agri-Fin Mobile, Reuters Mobile Light and IKSL, coordinated the dissemination of information around the crop cycle, with different information being provided around the relevant timings for each crop. This worked well since farmers were more likely to engage in responsive action when the information about good farming practices was provided at the appropriate time. Contextualizing information was also a key factor for success. The key informant interviews identified how this could entail translating the information into the farmers' local language, breaking down information into comprehensible pieces based on farmers' current knowledge and/or using local intermediaries, such as extension staff, to assist with interpretation. These intermediaries, sometimes

known as 'infomediaries', can access the information from the applications on behalf of farmers and/or share it and discuss it with them in a contextualized way. As shown in the Sesame Marketing Project, Community Knowledge Worker and ACRE initiatives, intermediaries can be local agricultural extension workers, lead farmers, or trusted agri-business owners, who, for example, also increase the credibility of the information provided.

1.3.4 Understanding constraints

The adaptation of delivery methods and communication approaches to cope with specific constraints was important. Many smallholder farmers have poor literacy and therefore interactive voice-based applications rather than SMS often worked better. The ability of farmers to interact with the applications and to tailor the information more towards their needs appeared to increase the likelihood of success, with the initial 'push' services in mKisan and Esoko, for example, giving way to more 'pull-' based applications. Complementary communication approaches linked to the applications, such as call centres and radio programmes, also appear to enhance success. Furthermore, the use of visual applications and video clips worked particularly well where the technology permitted. The following quote relates to one of the case studies which used tablets to convey information to sesame farmers in Tanzania.

> The ability to make extensive use of video and audio files is particularly suited to cultures with an oral tradition of learning and sharing of information, and for targeting users with low levels of literacy, which is a pervasive problem in rural communities in Sub-Saharan Africa.
> (Allan *et al.*, 2014, p. 13)

1.3.5 Weather and climate information

In terms of weather- and climate-related information in particular, several lessons could be learned. The review established that weather-related information was prioritized by farmers when they were asked about their information requirements from their mobile phones (Mittal *et al.*, 2010, Palmer, 2014). In most cases only basic short-term weather forecasts were communicated and there were no initiatives in which historical climate data was provided. The accuracy of the information and its localization was a concern particularly as weather patterns can vary significantly across even small geographical areas due to specific topographical variations (Pshenichnaya, 2012, Palmer, 2014). Indeed, some initiatives attempted to develop their own local weather stations to help improve forecasting. Farmerline developed a consortium under the umbrella of the TAHMO initiative (Trans African Hydrometereological Observatory) to develop a dense network of small weather stations to help to provide accurate weather information to 10,000 cocoa farmers on their mobile phones through voice messaging (Kaisaris, 2014). There was little evidence of weather and climate information being integrated into other agricultural information or being contextualized in any way so that it can be useful for decision-making purposes. There are therefore opportunities to improve the provision of this type of information using mobile applications. Looking ahead, there are technological advancements, such as GPS and GIS, which offer the possibility of providing more accurate, localized weather and historical climate data to smallholder farmers and of combining it with other agricultural information such as soil type and water management.

Overall, the review confirmed the demand for weather and climate information through mobile applications by smallholder farmers. It also highlighted some of the key factors for the success of mAgri initiatives reviewed, and these lessons learned were then drawn upon for the development of the two mobile applications. One of the lessons learned was the importance of co-developing some elements of the applications to ensure that they were meeting the needs of farmers and that any technological or socioeconomic constraints were being addressed. The significance of using accurate, localized information was also recognized. Furthermore, some of the more successful applications were designed to be visual and interactive wherever possible and to be delivered through intermediaries who could contextualize the information and therefore make it credible.

1.4 Development of the Mobile Applications

Using the lessons learned from the review wherever possible, two mobile applications (from now on referred to as apps) were developed to support different specific components of the PICSA approach. The two components were selected so that they might be sufficiently different, whilst also likely to lend themselves well to an app experience.

The first app developed (*see* Fig. 1.1) aims to present relevant climate information and provide a set of tools to support decision-making around the data. Within the app itself users can select data related to one of ten sites available for the northern region of Ghana, or automatically select their closest site using the device's GPS. There are then presented with four separate graphs available to view: seasonal rainfall totals, seasonal length totals, date of start of rains, and date of end of rains. The graphs can be interacted with by directly selecting points to view more information, or by inputting values into the line tool. The line tool is used to take an example of specific crop water requirements and automatically calculate how many years there have been totals both above and below this requirement. This information is then displayed as a line, as a probability (percentage) and as a '1 in x' visual block representation. Without the app, this probability is calculated by PICSA participants. The app is fully functional offline, with all data read from locally stored csv files and a map cached as a series of tile objects.

The second app (*see* Fig. 1.2) focuses on individual farmers and resource management. Within the app there are a series of time periods, activities, and associated budget considerations from which the farmer builds a picture of their seasonal budget. Farmers are guided to think about their net inputs, family labour hours, net outputs and produce consumed, which are then in turn quantified in terms of monetary value, time and/or resource consumption. For direct inputs and outputs, such as bags of seed and sacks of produce, farmers are prompted to input a number of units (be it bags, kilos or anything else) and unit costs. The app then automatically calculates running and total costs, the final cash balance, as well as net non-sale consumables and total family labour hours. Data within the

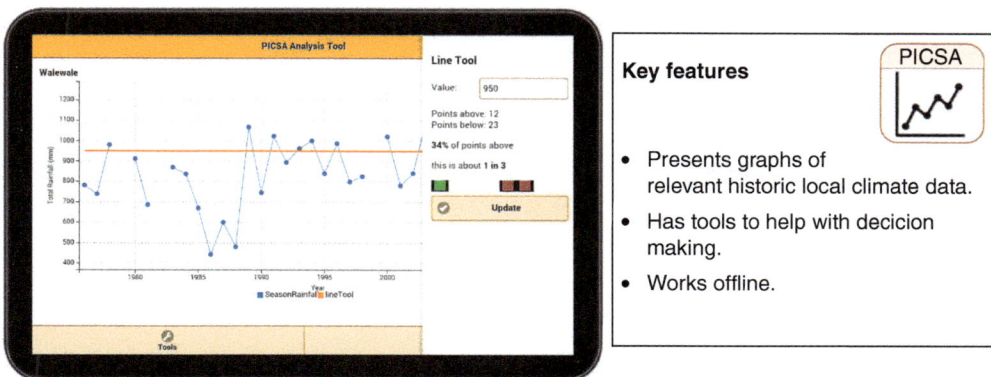

Fig. 1.1. Screenshot from the historical climate PICSA app.

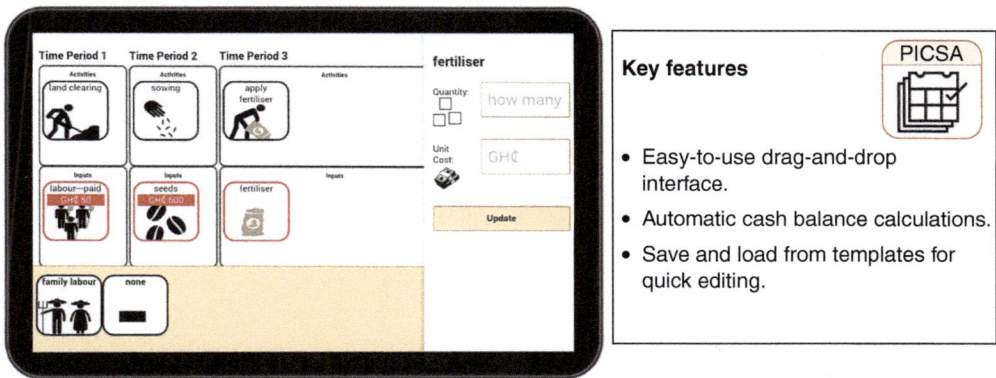

Fig. 1.2. Screenshot of the participatory budget PICSA app.

app is stored locally to avoid the need for internet access, and individual budgets can be loaded and modified either from those previously saved or templates included.

1.4.1 Proof of concept project

Based on the lessons learnt from the review, it was decided that the apps should be tested in the field with local farmers and extension officers during the early stages of development. This participatory approach to the app development was also in line with PICSA's core philosophy. The team initially hoped to generate feedback on the interest, requirements and ideas for PICSA apps, and how they might be used within the existing PICSA programme. Additional research questions included determining the extent to which demographic and socio-economic factors, such as gender or levels of literacy, might influence uptake and use of the apps. Finally, it was planned that some app development would take place directly in the field, providing the opportunity to enhance the app with local content as well as giving a sense of ownership to those providing input.

The proof-of-concept project took place with farmers across six communities in Northern Ghana, five of which had previously received PICSA training, and one that had not. Two extension officers visited each community with a set of eight tablets, and were accompanied by the lead app developer. Each community consisted of roughly 30–40 farmers, male and female, to whom the extension officers would first briefly introduce and demonstrate the apps, before splitting into groups so that the farmers could get hands-on experience of the apps for themselves. Figure 1.3 shows how several tablets were arranged in the centre of the meeting space to allow everyone to have a good view of the demonstrations. Each group was given a tablet to interact with and the extension officers moved between groups to support and encourage discussion. Final discussions were then held with all participants at the end.

1.4.2 Initial observations

The following section sets out the observations from the proof-of-concept project, which are based on notes taken during group and community discussions, and notes taken from discussions with extension officers after each session and at the end of all six training sessions.

All farmers responded very positively to the introduction of the apps, with many stating that it was something they could see immediate value in having. During demonstrations by the extension officers, farmers were actively engaged and provided lots of feedback when the extension staff started discussions. In particular, the historical climate app was recognized immediately by nearly all as a means to greatly facilitate the activities they had previously done with pen and paper; the graphs were identical to those on paper but you no longer needed to count or calculate, or manage multiple sheets. In the community that had not used PICSA before, both the climate app and paper graphs were presented and used for a short training session. At the end of the session a good amount of material had been covered, with the app appearing to take little additional time to understand, and the majority stated that they would prefer using the app over paper[1].

When the farmers were split into groups, it was interesting to see high levels of engagement with the apps. Despite the fact that very few farmers had ever used tablets before (only one farmer across all the communities owned a smartphone), most were keen to hold the tablet whilst trying to interact. At first it was not immediately obvious how to use all of the features, however after a very short period of time, users typically had mastered the basics of selection and navigation, as well as more advanced drag-and-drop and numeric inputs.

Even though the apps were written in English and literacy levels were reported by extension staff to be low, this did not have a noticeable difference on the farmers' abilities to use the apps. Comments were made on multiple occasions with regards to the participatory budget app that, although some people couldn't read or understand the words given to the drag-and-drop tiles, the symbols used made sense or at least could become familiar over time. It was also commented that those who are illiterate often struggle to hold and use pens to draw symbols anyway, and that interacting by touch is easier. The visual feedback generated by the app supported building the required utilization skills and also appeared to encourage people to participate and engage.

One further aspect that was particularly positive to see was that women in the communities seemed to be afforded equal access to the devices and within most of the sessions were actually the first to volunteer to come forward and experiment with the tablets (*see* Fig. 1.4). The extension officer claimed that within these communities it was common for women to be seen as 'innovators', who were happy to experiment and take risks whilst the men preferred to watch at first and avoid making mistakes.

One of the main talking points that arose within all the communities was how they might be afforded continued access to the applications and support once they are developed further. Access arrangements suggested included community ownership of tablets that might be stored in locations such as local schools, libraries or meeting spaces, as well as regular visits from extension or technical officers. It was also suggested that the community could assign one or two individuals to receive training to support the community instead of necessitating the availability of extension officers; it was claimed on numerous occasions that people are good at learning new things when they perceive value such as they had done, akin to the relatively recent widespread adoption of basic, or 'yam' phones. All communities also believed that the logistical issues of device charging could be resolved simply, as all either had some form of power (usually small solar generators) or knew of nearby places that could be visited on a weekly basis.

Fig. 1.3. Farmer groups using and engaging with the PICSA apps.

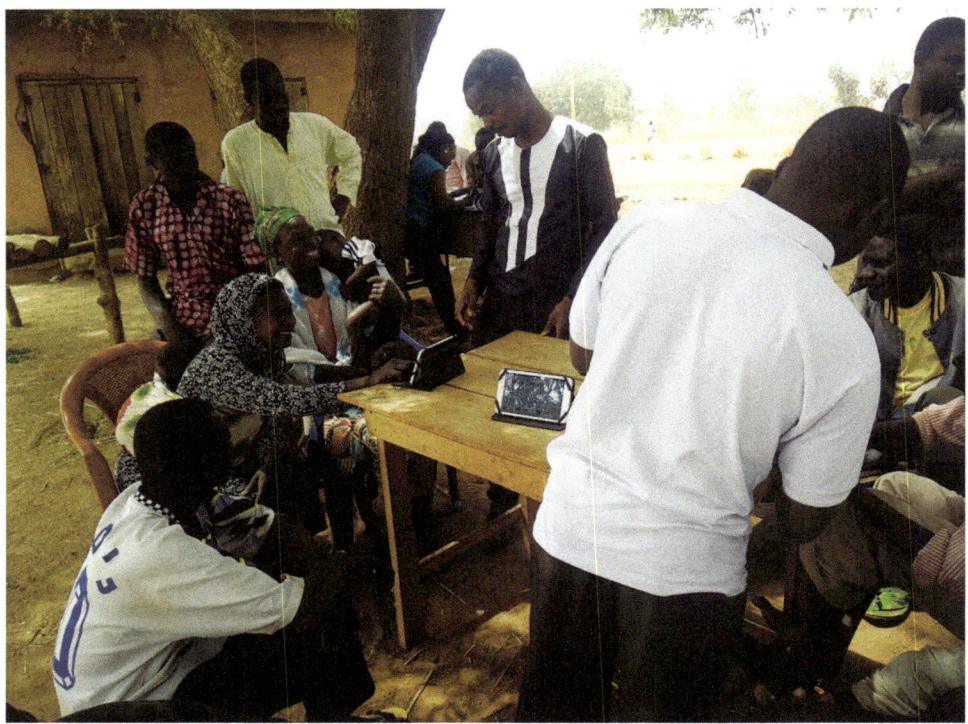

Fig. 1.4. Two women who were keen to be actively involved in learning to use the apps.

1.4.3 Main findings

Following the field-testing and subsequent discussions with farmers, extension officers, researchers and NGO staff, it appears that the apps do have the potential to facilitate and enhance areas of the PICSA approach. The historical climate app seemed to fit most naturally into existing training practices and with a few minor 'tweaks' could be considered ready to utilize as a tool for presenting graphs and calculating probabilities. The participatory budget app was interesting and definitely had its uses, however the fact that it took longer to master than pen/paper methods leads to the suggestion that in its current form it is not ready to integrate into the existing PICSA structure. The added value from the use of the app, such as the ability to quickly modify templates, explore different scenarios e.g. for a range of prices, and the rigour it encouraged when creating templates, should not be disregarded and could possibly fit into an additional training session within PICSA.

Establishing user needs and understanding constraints

The proof-of-concept project highlighted the importance of considering who exactly will be using the apps, establishing what they want to achieve from using them and understanding their particular constraints. The co-development of some of the parts of the apps in the field with farmers certainly helped with the process of developing a better understanding of these issues. As discussed in the content section below, it became clear that the farmers ideally required further information to make the apps useful and actionable from their perspective (information that is currently part of the PICSA approach but not through the app). It was also interesting to observe that traditional divides such as gender, literacy and digital skillsets did not appear to be a barrier to engagement with the apps. Overall, there was positive engagement with the technology. The interactivity of the apps was appreciated and appeared to heighten engagement and interest, and the visual nature of the apps allowed for illiterate farmers to use the apps with perhaps greater ease than with using pen and paper. The co-development of the apps benefited the process and led to improvements. To determine the exact contribution of this would require further research. The impact of some of the literacy and skillset constraints can be lessened through the use of well-designed applications that are tailored to farmers' needs.

The lack of internet access to provide updates and new information for the apps was another constraint that needs to be considered. Even in the larger nearby city of Tamale, accessing the internet was very slow and highly unreliable throughout. This highlighted the need for applications that retain full functionality offline, and alternative methods for updating and delivery such as distributing via USB sticks or Bluetooth. These in turn require additional thought around how to best ensure delivery and facilitation.

Content

The provision of location-specific climate information (which could be accessed using GPS) was appreciated by the farmers who, after an explanation from the extension workers about the source of the information (the National Meteorological Service), were assured about the credibility of the information. However, it was apparent that the app could have gone further in providing information that would support decision making. One question that arose was how to use the probability calculations to specifically determine which crops are particularly risky. It became clear that synthesising existing app information, such as water requirements and maturity length, with specific crop variety characteristics/requirements is needed to assist farmers in making choices around crop varieties based on climate information (again, something that is already part of the PICSA approach). The app could be developed further with additional graphs or tables comparing climate measurements with stored variety requirements and could help enhance the current methods for linking rainfall probabilities with crop varieties. In addition to the historical climate data, there were discussions about whether to include more recent climate information such as seasonal forecasts. Whilst this would be useful, it would require a significant amount of additional development and collaboration.

Intermediaries

The observations reveal how the farmers were keen to gain access to the apps without the need for the intermediaries such as extension workers. At the moment this is physically difficult because of the lack of devices (in this case, tablets) on which the community can obtain access to the apps. However, as smartphones become more readily available, this raises issues about the necessity of using intermediaries, who can contextualize the information and provide the valuable support to farmers in interpreting the information provided. This issue is discussed further in the future developments and further research section below.

1.4.4 Future developments and further research

In addition to the two apps presented, a number of potential future apps were discussed which could serve the PICSA approach well. In particular, an app specifically designed to support extension staff in their role could hold great benefit. This app should include key reference material like the PICSA manual (currently printed and carried around), frequently asked questions, tutorial videos as well as monitoring and evaluation support via digital surveys, templates and forms. The monitoring element would greatly help with tasks such as logging visits, locating communities, and accessing specific key information. As mentioned above, developments on the historical climate app will help to better integrate crop information for farmer decision making.

The development and testing of the apps raised several issues that need to be addressed through further research. Although the apps appeared to be well received by the farmers, it is uncertain to what the extent the use of the apps affected the success of the PICSA training compared to a more traditional approach that does not involve technology. Further research is required as to whether and how farmers engage differently when technology is used and to what extent gender, age, literacy and the novelty of the technology influence attendance at the sessions and engagement with the materials. Each tablet screen was visible to a small group of 6–10 farmers with one person operating, and it is uncertain how what may be perceived as a more individualistic approach could change the learning dynamics used within a participatory framework such as PICSA.

The proof of concept project raised questions about the best access and delivery models to be used. Although the extension workers and NGO staff acted as intermediaries to provide training on PICSA through the apps, the farmers wanted direct access to the apps themselves after the initial training. Although most farmers do not yet have smartphones, they may do in the near future and so further research is required to determine the advantages and disadvantages of farmers accessing the apps without intermediaries who can, for example, help contextualize the climate information.

There are also questions about how climate information is spread from farmer to farmer when the information is sourced from apps and how this may change decision-making processes by farmers, particularly if they have access to the app on their own mobiles. Furthermore, it is important to understand how the apps impact upon the role of the extension worker / community volunteer from their own perspective and performance.

have the potential to enhance the PICSA approach and therefore the sharing, analysis and use of climate information to support farmers in their decision making. The lessons learned from the review played an important part of the process of developing and testing these apps. Establishing user needs, understanding location specific constraints, developing accurate, localized and actionable content and the contextualization by intermediaries were all key considerations. Farmer engagement with the technology and with the content is complicated and there are a number of key questions that need to be explored as the apps are further developed for use on a larger scale in combination with PICSA. Nonetheless, the process has reinforced the principle that the apps need to be developed and refined in a process that puts the farmer at the centre.

Acknowledgements

The authors gratefully acknowledge: the support provided by the CGIAR Research Program on Climate Change, Agriculture and Food Security (CCAFS) which enabled the review reported in this chapter to be conducted; and the encouragement given by Philip Thornton and Wiebke Foerch.

Note

[1] This provides an indication of the interest that farmers showed in using the tablets as an information source but should not be considered as an objective assessment of their efficacy over paper.

1.5 Conclusion

The initial findings from the proof of concept project, which developed and tested two mobile applications, suggest that mobile applications do

References

Allan, C., Canales, C., Elibaraki, T., Knight, J. Marcheselli, M. *et al.* (2014) Elimsis – A mobile learning platform for strengthening extension services in Tanzania. Available at: http://elimsis.org/wp-content/uploads/2014/10/Brief_Farm-Africa-Final-Report_Brief.pdf (accessed 22 September 2016).

Caine, A., Dorward, P., Clarkson, G., Evans, N., Canales, C. and Stern, D. (2015) Review of mobile applications that involve the use of weather and climate information: their use and potential for smallholder farmers. *CCAFS Working Paper no. 150.* CGIAR Research Program on Climate Change, Agriculture and Food Security (CCAFS), Copenhagen, Denmark.

Dorward, P., Clarkson, G. and Stern, R. (2015) *Participatory Integrated Climate Services for Agriculture (PICSA): Field Manual.* Walker Institute, University of Reading, UK.

Glendenning, C.J. and Ficarelli, P.P. (2012) The relevance of content in ICT initiatives in Indian agriculture. *IFPRI Discussion Paper 1180*. IFPRI, Washington, DC.

Hellström, J. (2010) The innovative use of mobile applications in East Africa. *SIDA Review 2010: 12*. Swedish International Development Cooperation Agency, Stockholm. Available at: http://www.sida. se/English/publications/Publication_database/publications-by-year1/2010/june/the-innovative-use-of-mobile-applications-in-east-africa/ (accessed 6 March 2017).

Kaisaris, J. (2014) Farmerline launches TAHMO initiative. Available at: http://farmerline.co/blog/farmerline-launches-tahmo-initiative/ (accessed on 6 March 2017).

Kameswari, V.L.V., Kishore, D. and Gupta, V. (2011) ICTs for agricultural extension: a study in the Indian Himalayan region. *Electronic Journal of Information Systems in Developing Countries* 48(3): 1–12.

Masuki, K.F.G., Kamugisha, R., Mowo, J.G., Tanui, J., Tukahirwa, J., Mogoi, J. and Adera, E.O. (2010) The role of mobile phones in improving communication and information delivery for agricultural development: lessons from South Western Uganda. *Paper presented to Workshop at Makerere University*, Uganda, 22–23 March 2010. International Federation of Information Processing (IFIP) Technical Commission 9.

Mittal, S., Gandhi, S., and Tripathi, G. (2010) Socio-economic impact of mobile phones on Indian agriculture, *ICRIER Working Paper No. 246*, International Council for Research on International Economic Relations, New Delhi, India.

Palmer, T. (2014) What do Tanzanian farmers want from Agri-VAS? Available at: http://www.gsma.com/mobilefordevelopment/programme/magri/what-do-tanzanian-farmers-want-from-agri-vas (accessed 6 March 2017).

Pshenichnaya, N. (2012) Tigo and Technoserve pilot Tigo Kilimo service, first lessons learned. Available at: http://www.gsma.com/mobilefordevelopment/programme/magri/tigo-and-technoserve-pilot-tigo-kilimo-service-first-lessons-learned (accessed on 6 March 2017).

2 Smartphones Supporting Monitoring Functions: Experiences from Sweet Potato Vine Distribution in sub-Saharan Africa

Luka M. Wanjohi[1]*and Christopher A. Moturi[2]†

[1]*International Potato Center, Nairobi, Kenya;* [2]*School of Computing and Informatics, University of Nairobi, Nairobi, Kenya*

2.1 Introduction

Smartphones have been a trending technology for some time now, and their popularity is projected to continue increasing with time. For example, the International Telecommunications Union (ITU) put the number of active mobile broadband subscriptions in Africa at 177 million in 2015 (International Telecommunication Union, 2016). This is over four times the number in 2010. A number of studies have demonstrated clear advantages in using Information and Communication Technology (ICT) based systems over paper for data collection (Caeyers *et al.*, 2010; Zhang *et al.*, 2012). There are a good number of publications in the medical field on the use of new technologies for data collection in sub-Saharan Africa (Thriemer *et al.*, 2012; King *et al.*, 2014; van Dam *et al.*, 2017). Not as much has been published in the agricultural research field, save for anecdotal reports and web-based articles. This chapter sets out to explore the suitability of using a popular ICT tool in sub-Saharan Africa (SSA) – smartphones in the collection of monitoring data.

Malnutrition in Africa is one of the major public health problems; it claims the lives of millions of children every year. To address this challenge, the International Potato Center (CIP) is intensively working on promotion of production, consumption, marketing and processing of roots and tubers, and products based on these roots, with special focus on potatoes and sweet potato. Orange-fleshed sweet potato has proven to be an alternative and cheap means to address vitamin A deficiency (Webb Girard *et al.*, 2017). Selected farmers are provided with disease-free sweet potato vines for multiplication. The multiplied vines are then distributed to targeted households within the locality of the respective vine multiplier. A smartphone-based data collection tool was deployed to break barriers to access to information on vine dissemination, by providing an easy-to-use alternative for the stakeholders involved in the sweet potato value chain. Information on the vine multiplier's capacity to supply vines, their location and contact address is made available via online interactive maps, which get updated every season. Besides making quality seed information available, the tools

* E-mail: l.wanjohi@cgiar.org

† With additional contributions from: Temesgen F. Bocher, International Potato Center, Nairobi, Kenya.

were tested in monitoring and medium-level surveys and found to be more efficient in terms of time, money, and most importantly generating quality data.[2]

Pen-and-paper has been the standard method of collecting monitoring data. Details of beneficiaries of the sweet potato vines are also recorded using paper. Often, challenges associated with this method of data capture include delayed digitization, limited data capture and control options and ultimately delay in the dissemination of information to various stakeholders. Vine multipliers play an important role in bridging the seed availability gap. Unfortunately, without close monitoring, sweet potato seed handed down to a multiplier can easily become contaminated in the process of multiplication. During the period that these vines are being multiplied, a reliable monitoring system that enables immediate transmission of feedback to relevant stakeholders about the growing vines is required.

The objectives of this research were: (i) to identify challenges in the collection of monitoring data in the multiplication and dissemination of sweet potato vines; and (ii) to deploy a smartphone-based data collection solution that can address the challenges identified. This chapter documents to what extent smartphones can be used to bridge the monitoring data collection gaps. The lessons learned are transferable to other monitoring and evaluation (M&E) disciplines in agriculture in SSA. This study also documents important feedback to players in the ICT field who are developing smartphone-based applications and hardware devices for data collection in agricultural research in SSA.

2.2 ICT and Data Collection

Monitoring is the continuous process of systematic data collection on specified indicators to keep stakeholders up-to-date with the progress of a given indicator. Evaluation, on the other hand, is a process aimed at establishing the relevance, fulfilment of objectives, efficiency, effectiveness, impact and the sustainability of an ongoing or completed intervention (OECD-DAC, 2002). Data quality is at the heart of the success of any given M&E undertaking. The Food and

Agriculture Organization (FAO) traced weaknesses in M&E in agriculture to, amongst other things, underestimating complexities of data collection (Muller-Praefcke et al., 2010). The quality of monitoring data is a core component of a sound M&E system. Programmes must strive to make reasonable investments in M&E that ensure the collection of quality data and the integration of this data into programme management activities.

There is overwhelming evidence from various studies showing that the use of ICTs improves the quality of data collection. Caeyers et al. (2010) showed that computer-assisted personal interviewing (CAPI) virtually eliminates errors of missing variables, common with pen-and-paper interviewing (PAPI). Caviglia-Harris et al. (2012) reinforced this and found that CAPI increased the confidence in recorded data, and reduced data entry and cleaning time. Accurate and speedy processing of M&E data is all the more important when you have a crop with a short growing season, such as sweet potato. More importantly, especially for monitoring data, Caviglia-Harris et al. (2012) notes that CAPI makes it easy to collect survey metadata such as geographic coordinates, interview duration, enumerator identifications, etc. Metadata for a monitoring activity is critical in ensuring authenticity of data.

The need to process data accurately, and in a timely manner, is often competing with several other needs in any given programme, key among them being available financial and human resources. This is true not only for agricultural research and development but also in other disciplines. There are many studies documenting the use of CAPI in developed countries, but not for developing regions like SSA. This could easily be a result of the inequality in the amount of resources available for research in both places and the lack of reliable communication infrastructure in the latter case. CAPI hardware and software was not readily available five years ago for most researchers in the developing world. Caeyers et al. (2010) notes that the unit cost for a mobile personal computer used for a CAPI study in Zanzibar in 2009 was US$1800. Caviglia-Harris et al. (2012) makes note of a unit cost of US$3700 for a ruggedized laptop computer used in another CAPI survey in Latin America in 2009. These costs do not make the use of CAPI readily available to most research programmes, particularly so in the developing world. The emergence of Android smartphones provides an exciting possibility that the phone could be used as an affordable data collection tool. The smartphone has already displaced some commonplace items from their traditional roles: the traditional wristwatch, the alarm bell, the camera, the need to always have hard cash, to mention but a few. The growing popularity of Android smartphones worldwide coupled with their size and ability to accommodate complex functions creates a compelling case to consider them as possible tools for mainstream data collection. Smartphone software development has not lagged behind either, and readily available data collection software that can run on Android already exists developed to a fairly well refined level (Aanensen *et al.*, 2009; Anokwa *et al.*, 2009).

Karetsos *et al.* (2014) have reviewed the use and capability of the smartphone in agriculture and identified the potential use of smartphones and apps for the Android operating system as a promising solution to enable farmers' access to government information. Pongnumkul *et al.* (2015) have reviewed smartphone applications that utilize built-in sensors to provide agricultural solutions and identified four categories: farming applications; farm management applications; information system applications; and extension service applications. Bhavsar and Grijalva (2014) present a simple design framework to enable field-level workers such as agricultural extension workers to deliver outreach services in remote communities. Bobbie and Nsiah (2014) have developed a system for data capture in a monitoring and evaluation (M&E) environment. The prototype system uses Short Message Service (SMS) to reach out to stakeholders in the M&E processes for real-time impact assessment. Patodkar *et al.* (2015) developed an Android application that, based on sowing date of crop, reminds farmers of application of fertilizer, herbicide as per schedule, pesticide for diseases and weather alerts if particular crops exceed favourable temperature ranges. An interactive smartphone app by Vellidis *et al.* (2014) provides notifications to cotton farmers when the root-zone plant-available soil water deficit exceeds 50% indicating that irrigation is recommended. Roberts and McIntosh (2012)

have shown the potential benefits of using smartphones and tablets as a major extension information tool by advisors, farm merchants and farmers.

Some of the challenges of implementing smartphone-based data collection are the need for more enumerator training and the limited supply of smartphones. Impediments to adoption and diffusion of technology include the lack of awareness, low literacy, infrastructure deficiencies such lack of electricity to charge the smartphones, language and cultural barriers, low e-inclusivity, and the need to cater for the special needs of some users (Maumbe and Okello, 2010; Aldhaban, 2012; White, 2013). Lessons from field studies have shown that the use of smartphones in data collection can improve data quality, integration and speedy dissemination of results, and visualization of the same using charts and graphs for enhanced data monitoring and reporting (Brennan, 2014). Solutions include Open Data Kit (ODK) software, a system that digitizes data for analysis, allows for remote monitoring of the collection progress, and facilitates the gathering of data, eliminating the need for paper surveys (Hartung *et al.*, 2010). This significantly reduces survey turnaround times (Jeffrey-Coker *et al.*, 2010). ODK has the potential for a profound impact on the future of data gathering, particularly in development applications where locations may be remote and budgets tight, yet where mobile phone use is rapidly increasing with the expansion of service coverage.

2.3 Methodology

This study deployed electronic forms on smartphones to collect monitoring data. The electronic forms were modelled after existing paper forms used in monitoring activities of sweet potato vine multipliers in SSA. Existing monitoring staff were trained on how to use the new tools.

2.3.1 Case study design

To maximize richness and accuracy of data, as well as transferability of the findings, a case study was carried out in nine different countries in SSA. The nine countries were: Burkina Faso, Ghana, Nigeria, Kenya, Ethiopia, Uganda, Rwanda, Tanzania and Mozambique. In all these countries, individual vine multipliers were identified and provided with disease-free sweet potato vines. Special training was provided on how to multiply sweet potato vines and technical backstopping was provided throughout the multiplication period. CIP collaborated with government agencies to identify suitable farmers to work as sweet potato vine multipliers. Technical backstopping support was provided jointly by CIP agronomists, government officers and other development partners in the respective areas. The smartphone-based M&E system was employed to track the process of growing the vines and distribution of the same.

Previously, paper-based tools were used for collecting monitoring data for such projects. These tools were:

- Paper form to register a vine multiplier.
- Paper form to register the field characteristics of each vine multiplier.
- Paper form for recording beneficiaries of vines when dissemination begins.

Monitoring of the multipliers' fields is a regular activity carried out by government extension officers and in other instances by CIP agronomists. Extension officers are trained agricultural personnel who offer technical backstopping services to farmers.

The sweet potato vine distribution monitoring data collection system was setup to help:

- Maintain an in-country geo-referenced database of all sweet potato vine multipliers.
- Monitor the quality of sweet potato vines being produced for distribution.
- Maintain an in-country database of all households receiving sweet potato vines.

Sweet potato vine multipliers would be recruited at the beginning of every planting season. These multipliers were categorized into three tiers, depending on their source of planting material and the level of technical backstopping offered by CIP. The project manager maintained the database of all the vine multipliers and this would be updated annually.

Paper-based data collection tools would be developed at the project level in consultation with relevant colleagues across CIP. Once a given tool is ready, it would be printed out and tested in the field. Any changes required would be noted during the testing process and these would get incorporated into the next revision. After implementing all major corrections, the tool would be ready for data collection in the field (see Fig. 2.1). At the commencement of field data collection, a data entry tool would be developed in tandem to digitize the data once it comes from the field. The choice of data entry software was largely left to the relevant M&E staff to decide. The tool would then undergo iterative testing with real data coming from the field. Data-entry clerks would then be hired and trained on how to use to the tool before they begin data entry. Equivalent electronic forms were developed from these existing paper forms. The electronic forms were loaded on smartphones which were then handed over to various field staff. The field staff were trained on how to use the smartphone-based M&E tools before the start of the sweet potato vine multiplication season.

At the onset of the season, the field staff registered each vine multiplier contracted by CIP using the electronic registration form. This form collected, among other standard sets of personal identification information, the geographic coordinates of the vine multiplier and a photograph. The form also registered the details of a multiplier's field (e.g., the size, the location, the expected output, etc). Every two to three weeks the field staff were expected to visit the vine multiplier's fields to check on progress and offer any technical assistance that may be necessary. These visits were logged using an electronic monitoring form including date of visit, picture of the sweet potato crop progress in the field, any challenges being experienced by multiplier including a picture of the field, etc. Once the vines being multiplied were ready for distribution to final beneficiary households, the vine multiplier would record details of each beneficiary coming to collect vines on a sheet of paper. Completed forms would then be handed over to the field staff responsible at regular intervals. The field staff would digitize this data using an equivalent electronic form on their smartphone. Distribution of vines could either be at the multiplier's farm or it could be at a pre-publicized

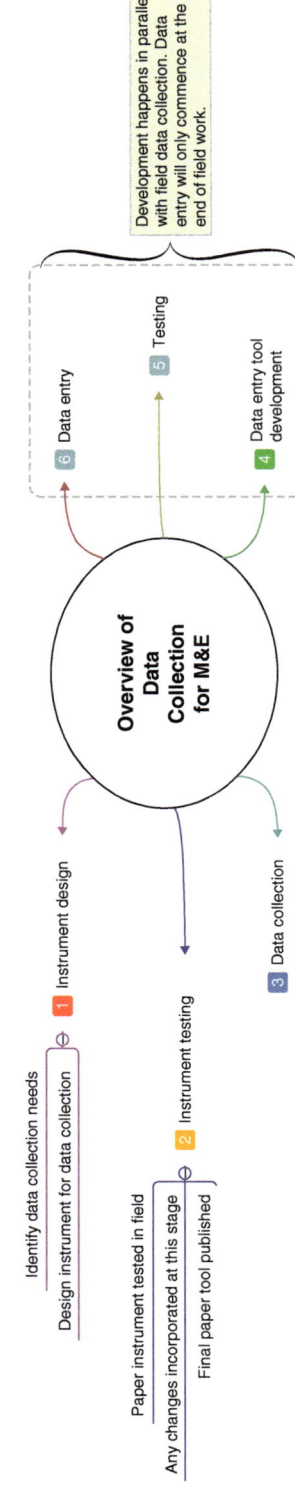

Fig. 2.1. Pen-and-paper data collection overview.

event at the local market or another public place (e.g., church, school, etc).

The smartphone pilot was carried out over a period of two years, starting with Nigeria in May 2014 and gradually expanding to other countries in SSA every sweet potato growing-season. All the data entered were stored in a secure database accessible over the internet. Transmission of the data was via cellular networks. Key learning points were documented throughout the process, and changes made to the original design to respond to unaccounted-for real-life situations. Throughout the whole pilot, limitations and strengths of this system were noted and compared to previous monitoring data collection exercises using pen-and-paper.

2.3.2 Data collection and analysis

Field visits were conducted, in the company of the respective field staff. These helped in understanding the data collection processes and appreciating the context in which the data were collected. The field staff were interviewed during these field visits wherever necessary to clarify on workflows. Extensive interviews were conducted with all participants before, during and after the field visits to ensure that no important detail about the monitoring data collection setup was left out. The major challenges identified with the pen-and-paper monitoring system were as follows.

- It took a long time between data collection and digitization of the same for any meaningful feedback for project management purposes.
- Important monitoring data sometimes never got to achieve its intended purpose, as paper forms got lost or were never digitized. This happened several times because data were being collected by extension officers who in turn had to arrange for transmission of the data to the CIP offices.
- Errors would get introduced between collecting and digitizing data. This would be due to the data being digitized by different persons from the ones who originally collected it.

- In the paper-based system, different data types would be stored on different devices: photos on the camera, GPS coordinates on the GPS devices, and text data on the paper form. Aggregating all the different data did not always happen in a timely manner. In addition, errors often got introduced during the aggregation (e.g., a photograph would end up being matched with the wrong personal identification information, etc.).
- Management of geographic location data was a challenge to many people who have no formal training in the subject. Most staff would normally read the GPS coordinates from the device and manually write the coordinates on a sheet of paper with the multiplier names. Sometimes they failed to indicate whether a given GPS reading was in the north or south; other times readings in degree decimals were mixed up with readings in degree minutes or degree minutes seconds. This increased the amount of data cleaning required before such data were of any use.
- Handling dates is a challenge especially when you are working across different countries and cultures. People do not always stick to the same convention of reporting, e.g. DDMMYYYY for day/month/year respectively. There is always a high risk of using the day and month interchangeably, resulting in bad data.
- The old system was open to fraud. There was no way of ensuring that the extension officers are actually visiting the multipliers. Reports could be filed from the comfort of an office and sent in without actual field visits.

2.4 Pilot Deployment

To address the above challenges, Android smartphones were selected for the pilot implementation. Android is an open source mobile operating system developed by Google. Currently it is the most popular mobile operating system (Chau et al., 2016). Open Data Kit (ODK) was used to implement mobile data collection on the Android platform. ODK is an open source suite of technologies developed at Washington State

University, with three main components: form builder, ODK collect, and ODK aggregate. The form builder is used for programming your electronic data collection form. XLSForm standard makes it possible to author forms using a spreadsheet application like MS Excel. ODKCollect is a mobile app that runs on the Android smartphone and displays these forms to users during data collection. ODKAggregate is a ready-to-deploy server and data repository that provides blank forms to ODKCollect, accepts finalized forms from ODKCollect and allows exporting of uploaded data into various formats (e.g., CSV files). To collect data, no active data connection is required; however, the Android smartphone must be connected to the internet. An ODK aggregate server, Azizi, run by the International Livestock Research Institute (ILRI) was made available for use at no cost at the start of the pilot.

Nine Google Nexus 5 smartphones were purchased in the first phase of the pilot in Nigeria. The pilot commenced by attempting to register vine multipliers in six different states in Nigeria. Over time, additional hardware has been purchased throughout the nine countries. Each piece of hardware acquired after the first Nexus devices was informed by experiences in the field. Each device was set up with a mobile data connection. Three monitoring forms were authored using the XLSForm standard: Vine Multiplier Registration Form, Vine Multiplier Monitoring Form and Vine Beneficiary Registration Form. Different forms were authored for each of the different countries. Initially, all forms were authored and tested in Nigeria. Afterwards, intensive in-country training on ODK form development was delivered to selected staff in the pilot locations. The staff trained were mainly those in charge of monitoring and evaluation activities in the respective countries. They were expected, at the very least, to be able to modify an existing XLSForm to suit country-specific conditions such as administrative locations, local units of measurements, etc.

Field data collection staff were trained on how to collect data using the electronic forms. Data collection staff included CIP agronomists and government extension officers. In other instances, we had staff from partner organizations, but mostly with an agronomic background. Training was done in country, with all participants gathered in one place. This would be followed by a hands-on exercise in a real vine multiplier field. In every country, the CIP staff in charge of monitoring and evaluation was responsible for all coordination and planning.

2.4.1 Key outcomes

As already confirmed by a number of other studies, smartphone-based data collection improves the quality of the data drastically. The key outcomes over the period of the pilot are summarized below.

Improved management of in-country vine multipliers databases

There was a significant reduction in the amount of time it took to process monitoring data (i.e., from data collection to digitization and dissemination). By August 2016, the system had been used to register 326 vine multipliers in the nine different countries. 253 of these multipliers were male and 73 were female. 250 monitoring visits have been logged using the smartphone system in Burkina Faso, Ghana, Nigeria and Kenya. We do not have any evidence to report on how these have influenced programme management, but it is the first time that monitoring data is being made readily available to project management as soon as it is collected. Electronic data collection made it easy to deploy standardized forms across different countries. As a result, data cleaning efforts post-collection were minimal; the data having a similar schema meant a huge time saving in the amount of time previously spent processing datasets individually due to differences in the way they were collected.

Improved data integrity

Overall, there was a great improvement in the accuracy and consistency of our monitoring database across the region. Electronic data collection made it easier to use uniform data coding schemes and improved quality controls across different countries and projects. As an example, all location data are now usable since there is no room for collecting these data using different coordinate formats like before. Similarly, all dates are uniformly recorded. It also became easier to record the names of vine multipliers in a consistent way across different countries without mixing up, for example, the surname and the first names because of differences across cultures. Field staff reported that the monitoring metadata reported (such as the data collection time) made them keener on ensuring they are providing farmers with backstopping visits as required.

Improved data management of multiple data types

Using the electronic data collection forms enabled our enumerators to collect and store text data, pictures and location information all in a single record. This had the obvious advantage of making it easy to refer back to any of these data whenever necessary. It also reduced significantly the risk that any of these different data entities ended up with the wrong labels or getting lost all together. Several projects participating in the study did not have dedicated resources for ICT-related support. Having a system that project staff could work with easily, despite having varying levels of knowledge on ICTs, was a big advantage.

The system was easy to replicate across the different countries where we work. Within the two years of the pilot, we were able to scale up capacity to develop XLSForms across the SSA region. Key staff members could work with an already developed form and customize it to suite in-country needs. The ability to have in-country capacity to do basic technical backstopping was critical in guaranteeing sustainability. In Mozambique, Ghana, Nigeria and Burkina Faso, independent trainings on XLSForm development and collection of monitoring data using ODK have been conducted and these were fully facilitated by trainees from the initial pilot phase. The original XLSForms underwent a lot of improvements based on feedback from the field. Every time a new version of the form was released it was circulated to the persons in charge of each country who were then responsible for adapting it and updating the same for all of their data collection staff. Across all countries we maintained the same variable names, with individual projects being at liberty to translate questions into a local language where it was deemed necessary. Being able to achieve this type of

standardization across different countries with very different cultures and languages was an important achievement for sweet potato research.

2.4.2 Challenges encountered

The above observations are not new. They have been confirmed by previous studies (Caeyers et al., 2010; Caviglia-Harris et al., 2012), albeit in a survey set-up. Achieving these in a monitoring set-up however was very important for our work. Most of the work carried out in CAPI before has been evaluated during major household surveys. A survey, being a one-off activity, makes it easy to take advantage of concentrated resources to manage for the duration. It is likely that enumerators of choice can be enlisted and this makes the subsequent training easier. For the short duration that a survey occurs it is possible to provide high-level field data collection supervision and support.

It is important to note that data-collection field staff consisted of agronomists and extension workers because they are professionally trained to provide the technical backstopping support required by vine multipliers. They had diverse backgrounds; age, gender, education and field experience. Some had never used smartphones before. One of the critical aspects in providing remote support is that the person being supported is able to describe any challenge that she/he is facing in the field with a given system as accurately as possible. This was not the case with every one of our field staff. Some of the challenges reported from the field would be related to not being able to navigate through the Android system, not being able to turn on location services to take the GPS coordinates or not being able to tell whether a given device had active internet connectivity or not. Continuous field training, remote support via phone calls, support via remote control applications such as TeamViewer and the use of WhatsApp groups organized by country, have seen the majority of the field staff get up to speed with using the system.

The ODK aggregate platform chosen at the start of the pilot could only provide data exports as comma separated files (CSV). Many researchers found this inadequate since they had to regenerate variable labels and any accompanying value labels for further analysis using statistical packages of their choice. Furthermore, as with most open source applications, the user interface of the basic ODK aggregate server is not refined and most users without a background in ICT found this difficult to use. Later on in the pilot the issue of usability of the data export interface was resolved by using a paid ODK hosting service by a company called ONA.[3] However, users are still keen on having more data export capabilities out of the box (i.e., the ability to export data from the ODK server into an easy-to-use format for a number of the major statistical analysis packages).

A few project management staff found useful the ability to access vine multiplier monitoring visit logs, as soon as they are uploaded. This was most important for projects operating over vast geographical locations such as a project that was operating across six different states in Nigeria. The logistical efforts involved in getting the paper data back to the office from the field can be overwhelming especially for a project with a small monitoring budget. Automated time stamps and location data collected at the time of data entry played an important role in increasing the validity of monitoring data collected. These helped project managers monitor their field officers.

The initial smartphone model selected at the start of the pilot (2013 Google Nexus 5), was gradually replaced with seven-inch Android tablets. The small screen size of the Nexus 5 (five inches) made it prone to misuse, as it was easy to carry around even during non-working hours and at the same time a number of staff involved in collecting data complained of difficulties in typing on the small screen. Over the two-year period, we tried out a number of different devices from various manufacturers. Most of these averaged between US$100 to US$300. Major hardware-related issues reported include poor screen legibility for some devices while out in the field and for others that it would take a long time before getting a GPS reading with an accuracy of at least 10 m. Over time we settled on Samsung tablets with a screen size of approximately seven inches, available for US$200 to US$300.[3]

It was not easy to use the electronic forms for data collection under strenuous conditions such as registering vine beneficiaries in a queue during dissemination. It took a long time to enter beneficiary data using a smartphone as opposed to using the traditional pen-and-paper. Vine beneficiaries have to queue up for their details to be recorded before they can proceed to pick up their vines. To address the digitization challenge for this particular set-up, data were recorded on paper and later all forms were scanned using a smartphone and uploaded onto an online server. We used a free application from the Google Play Store (CamScanner) to scan beneficiary data into portable document file format (PDF) before uploading them online. These forms were downloaded regularly and printed out in readiness for digitization using the Census and Survey Processing System (CSPro). CSPro is a public-domain software package developed by the United States Census Bureau for entering, editing, tabulating, and disseminating census and survey data.

2.5 Conclusion

The last few years have seen the use of smartphones, for data collection and for agricultural research and development work, gather significant momentum the world over. This study sought to establish the extent to which smartphones can be used to support monitoring activities for agricultural research and development activities in SSA. An ODK-based data collection solution was deployed to address challenges identified using paper-based monitoring data-collection methods for sweet potato research. The pilot has seen a significant improvement in the management of sweet potato vine multiplier databases across countries. It has been easier to promote the use of standardized data collection tools across the different countries where the pilot has been running, and this has seen an overall improvement of various data management processes.

Various challenges have been encountered along the way. At the start of the pilot in 2014, the cost of hardware was prohibitive for most participants who were invited to participate in the study. Although the prices of most 7 inch tablets that were used in this study have reduced by more than half to date, not every project manager will readily invest in one. A significant investment has gone into setting up the data collection forms, platforms and training. Scaling-up the pilot across countries has been successful largely due to capacity building among in-country monitoring and evaluation staff.

Sweet potato planting material multiplication has the potential to grow into a profitable business if the demand for orange-fleshed sweet potato grows across the SSA region. Given the popularity of Android smartphones in this region, some future plans would include providing vine multipliers with an ODK-based self-registration platform. Such a platform would be managed by seed regulation agencies of a given country. With the improved quality of the data coming in from the field and the standardization in place, immediate efforts can be geared toward the development of a data portal to host near real-time summarized status of sweet potato in SSA.

Notes

[1] Our thanks to Sweetpotato Action for Security and Health in Africa (SASHA) project led by the International Potato Center for supporting this study.
[2] See in particular: http://www.sweetpotatoknowledge.org/files/sasha-brief-2016-the-monitoring-learning-and-evaluation-community-of-practice-embraces-smartphones-to-support-monitoring-activities/ (accessed May 2017).
[3] ONA is a private technology company that provides commercial data hosting services for data collected using Open Data Kit. The company is based in Nairobi and Washington DC (www.ona.io).

References

Aanensen, D.M., Huntley, D.M., Feil, E.J., al-Own, F. and Spratt, B.G. (2009) EpiCollect: linking smartphones to web applications for epidemiology, ecology and community data collection. *PLoS ONE* 4(9), e6968.

Aldhaban, F. (2012) Exploring the adoption of smartphone technology: literature review. In: *Technology Management for Emerging Technologies, Proceedings of PICMET'12,* IEEE. Piscataway, New Jersey, USA, pp. 2758–2770.

Anokwa, Y., Hartung, C., Brunette, W., Borriello, G. and Lerer, A. (2009) Open source data collection in the developing world. *Computer* 42(10): 97–99.

Bhavsar, M. and Grijalva, K. (2014) From paper to mobile: design considerations for field level worker programs. In: Niang, I., Scharff, C. and Wamala, C. (eds) *Proceedings of the 4th International Conference on M4D Mobile Communication for Development, April 2014, Dakar, Senegal.* Karlstad University Studies, Karlstad, Sweden, pp. 255–259.

Bobbie, P.O., and Nsiah, J.R.O. (2014) Design and development of a system for data capture in a monitoring and evaluation environment. In: Arabnia, H.R., Deligiannidis, L., Solo, A.M.G. and Tinetti, F.G. (eds) *Proceedings of the International Conference on Internet Computing (ICOMP). July 2014, Las Vegas, USA.* CSREA Press, Athens, Georgia, USA, pp. 36–42.

Brennan, B. (2014) e-Mobile data collection to improve data quality and speed of results: lessons from Angola. Presented at 142nd APHA Annual Meeting and Exposition, November 2014, New Orleans, USA. Available at: http://www.africairs.net/2014/12/mobile-data-collection-improves-reporting-irs/ (accessed September 2017).

Caeyers, B., Chalmers, N. and De Weerdt, J. (2010) A comparison of CAPI and PAPI through a randomized field experiment. Available at: https://ssrn.com/abstract=1756224 (accessed May 2017).

Caviglia-Harris, J., Hall, S., Mulllan, K., Macintyre, C., Bauch, S.C. *et al.* (2012) Improving household surveys through computer-assisted data collection use of touch-screen laptops in challenging environments. *Field Methods* 24(1): 74–94.

Chau, M., Govindaraj, N., Reith, R. and Nagamine, K. (2016) IDC: smartphone OS market share. Available at: http://www.idc.com/promo/smartphone-market-share/os (accessed May 2017).

Hartung, C., Lerer, A., Anokwa, Y., Tseng, C., Brunette, W. and Borriello, G. (2010). Open data kit: tools to build information services for developing regions. In: *Proceedings of the 4th ACM/IEEE International Conference on Information and Communication Technologies and Development.* December 2010, London, UK. ACM, New York, USA. doi: 10.1145/2369220.2369236

International Telecommunication Union (2016) ICT statistics. Available at: http://www.itu.int/en/ITU-D/Statistics/Pages/default.aspx (accessed May 2017).

Jeffrey-Coker, F., Basinger, M. and Modi, V. (2010) Open data kit: implications for the use of smartphone software technology for questionnaire studies in international development. Columbia University Mechanical Engineering Department. Available at: http://qsel.columbia.edu/assets/uploads/blog/2013/06/Open-Data-Kit-Review-Article.pdf (accessed May 2017).

Karetsos, S., Costopoulou, C., and Sideridis, A. (2014) Developing a smartphone app for m-government in agriculture. *Journal of Agricultural Informatics* 5(1): 1–8.

King, C., Hall, J., Banda, M., Beard, J., Bird, J. *et al.* (2014) Electronic data capture in a rural African setting: evaluating experiences with different systems in Malawi. *Global Health Action* 7(1), doi: 10.3402/gha.v7.25878

Maumbe, B.M. and Okello, J. (2010) Uses of information and communication technology (ICT) in agriculture and rural development in sub-Saharan Africa: Experiences from South Africa and Kenya. *International Journal of ICT Research and Development in Africa* 1(1): 1–22.

Muller-Praefcke, D., Lai, K.C. and Sorrenson, W. (2010) *The Use of Monitoring and Evaluation in Agriculture and Rural Development Projects*. Food and Agriculture Organization, Rome, Italy. 63 pp.

OECD-DAC (2002) *Glossary of Key Terms in Evaluation and Results-based Management*. OECD, Paris.

Patodkar, V., Simant, S., ShubhamSharma, C.S. and Godse, S. (2015) e-Agro android application (integrated farming management systems). Available at: http://pnrsolution.org/Datacenter/Vol3/Issue1/47.pdf (accessed May 2017).

Pongnumkul, S., Chaovalit, P. and Surasvadi, N. (2015) Applications of smartphone-based sensors in agriculture: a systematic review of research. *Journal of Sensors* article ID 195308, 18 pp. doi:10.1155/2015/195308.

Roberts, K. and McIntosh, G. (2012) Use of mobile devices in extension and agricultural production: a case study. In: Yunusa, I. (ed.) *Capturing Opportunities and Overcoming Obstacles in Australian Agronomy, Proceedings of 16th Australian Agronomy Conference 2012.* Available at: http://www.regional.org.au/au/asa/2012/precision-agriculture/8224_robertsk.htm (accessed 29 May 2017).

Thriemer, K., Ley, B., Ame, S.M., Puri, M.K., Hashim, R. *et al.* (2012) Replacing paper data collection forms with electronic data entry in the field: findings from a study of community-acquired bloodstream infections in Pemba, Zanzibar. *BMC Research Notes* 5(1): 113.

van Dam J., Omondi Onyango K., Midamba B., Nele Groosman, N., Hooper, N. *et al.* (2017) Open-source mobile digital platform for clinical trial data collection in low-resource settings. *BMJ Innovations* 3(1), 26–31. doi:10.1136/bmjinnov-2016-000164

Vellidis, G., Liakos, V., Perry, C., Tucker, M., Collins, G. *et al.* (2014) A smartphone app for scheduling irrigation on cotton. In: Boyd, S., Huffman, M. and Robertson, B. (eds) *Proceedings of the 2014 Beltwide Cotton Conference, New Orleans, LA*, National Cotton Council, Memphis, TN, USA. Paper 15551.

Webb Girard, A., Grant, F., Watkinson, M., Okaku, H.S., Wanjala, R. *et al.* (2017) Promotion of orange-fleshed sweet potato increased vitamin A intakes and reduced the odds of low retinol-binding protein among postpartum Kenyan women. *The Journal of Nutrition.* doi:10.3945/jn.116.236406

White, E. (2013) The smartphone revolution: can technology benefit data collection in rural Ethiopia? Report of World Food Prize Foundation, International Livestock Research Institute, Addis Ababa, Ethiopia.

Zhang, S., Wu, Q., van Velthoven, M.H., Chen, L., Car, J., Rudan, I. and Scherpbier, R.W. (2012) Smartphone versus pen-and-paper data collection of infant feeding practices in rural China. *Journal of Medical Internet Research* 14(5): e119.

3 Customized Information Delivery for Dryland Farmers

Amit Chakravarty,* V.V. Sumanthkumar and Mukund D. Patil

*International Crops Research Institute for the Semi-Arid Tropics (ICRISAT),
Hyderabad, India*

3.1 Introduction

In India the number of mobile network subscribers had reached 969.89 million in March 2015. Of this around 43% (414.18 million) were rural subscribers (TRAI 2015). A rural population of 833 million (according to the 2011 census of India) implies that every second person in rural India owns a mobile phone. In contrast, only 12 out of 100 people in rural areas have access to the internet (TRAI, 2015). These trends are mirrored across the world. According to the *World Development Report*:

> On average, 8 in 10 individuals in the developing world own a mobile phone, and the number is steadily rising. Even among the bottom fifth of the population nearly 70% own a mobile phone. But internet adoption lags behind considerably: only 31% of the population in developing countries had access in 2014.
>
> (World Bank, 2016, p. 6)

Given these statistics and the extremely low cost of transmitting information over a mobile network, the mobile phone emerged as the number one choice to overcome the information deficit among the farming community. Mobile phones have the potential to deliver information to the farmer on demand rather than the farmer spending time in searching and procuring information. During peak agriculture seasons farmers cannot spend time procuring information without affecting their farming activities negatively. There are information search costs associated with each step in the chain of agricultural activities. These costs can be as high as 70% of all transaction costs, while transaction costs themselves are 15% of all costs incurred (Banerjee, 2013).

Thus, lack of information is considered to be a major impediment to improving farmers' livelihoods. To reach out to farmers with information about new tools, technologies and agricultural practices, the Government of India has in place a large workforce of extension agents. The agents reach out to farmers through various means, such as: one-on-one contact; farmer field days; farmer schools; wall writings; distribution of pamphlets/leaflets; and various other means to disseminate information and provide advice. In addition to the extension network, other sources of information for the farmer are peer-to-peer sharing, traders/intermediaries and input service providers, especially for seeds and fertilizers. Agriculture extension services, as they are currently structured, have several deficiencies:

- The coverage and effectiveness of extension services in India is very poor where the ratio of extension agents to farmers is 1:3000 (Vodafone, 2016), while some estimates put

---

* Corresponding author e-mail: a.chakravarty@cgiar.org

it even higher at 1:5000 (Mukherjee and Maity, 2015).

- The extension services are supply driven rather than demand driven.
- Farmers in remote rural areas or small and marginal farmers may not be reached due to difficulties in transportation and high cost of delivering information in person (Cole and Fernando, 2012).
- Extension agents may not be able to deliver timely information, say, when a farmer is facing uncertain weather or an unfamiliar pest infestation (Cole and Fernando, 2012).

The above-mentioned limitations interact and reinforce each other to limit the information reaching the farmers, even when they are reached by the extension network. This creates a large potential for mobile-phone-enabled extension systems to improve farmers' access to knowledge.

India has a long history of use of Information and Communication Technologies (ICTs) for agriculture. Some of the early pioneers were Warana Wired village (launched in 1998), Gyandoot (launched in 2000), Nokia Life (launched in 2009), Reuters Market Light (launched in 2007), e-Sagu (launched in 2004), e-Krishi, e-Choupal and iKisan. These initiatives were not limited to agriculture, they also provided information on other subjects like education, health, entertainment, provision of government services like birth/death certificates, copies of land titles, information on government schemes, government subsidies, and a variety of other information and services. However, the initiatives mentioned above are more focused on providing information and services related to agriculture only. There is a mix of government-led projects, non-governmental organization (NGO) led projects, as well as private-sector-driven projects. In terms of information delivery channels, the primary channels are: (i) operator-mediated computer kiosk; (ii) telephony (call centres and mobile phones); (iii) web portals; and (iv) different combinations of first three channels.

This chapter aims to demonstrate that compared to a generic information dissemination model, a targeted information dissemination model, which provides farm-specific information

to farmers based on their profile, helps farmers raise their productivity and incomes.

3.2 Uses of Digital Technologies in Agriculture

The primary uses of digital technology in agriculture are:

- *Agriculture extension* – to increase on-farm productivity by disseminating information on new tools, technologies, varieties, improved practices, weather information and pest and disease attack warnings.
- *Market transparency* – to provide market information enabling farmers to overcome information asymmetries and reduce dependence on market intermediaries.
- *Efficient logistics* – to connect farmers directly to buyers, enabling farmers to realize a higher proportion of the final price by cutting out intermediaries.
- *Financial inclusion and insurance* – to help farmers access finance and affordable insurance products to better manage risk.

The importance given to each of the above categories by farmers varies by region and type of crop grown. Some studies have found farmers prioritize market prices while others find weather is a top priority for farmers. In a small field study carried out in Bundi District, Rajasthan, a majority (57%) used ICT tools for getting market prices, followed by production practices (55%), plant protection measures (52%) and weather information (49%). The top four perceived benefits were direct access to information, wider subject coverage, minimized time/distance barriers and reduction in transaction costs (Dhaka and Chayal, 2010). In another study conducted with 1,100 farmers in one district each in the states of Haryana and Bihar in India, weather information was valued the most over all other types of information (Mittal, 2016). Another study (Kameswari *et al*, 2011) in the Himalayan State of Uttarakhand, found that information-seeking behaviour was determined by the cropping pattern. In villages where agriculture was profitable, farmers sought diverse kinds of information, whereas in villages where traditional cropping patterns were followed, farmers were more reliant on input dealers/intermediaries rather than seeking new information.

3.3 Utility of Current Information Services

Despite the number of services and service providers for providing information to farmers, whether it be for crop advisory, weather or markets, there are a number of outstanding concerns:

- *Use of SMS.* Given the low levels of literacy in rural India, there are concerns whether text messages are the most appropriate form of providing information. To overcome the literacy issue some service providers use voice messages.
- *Use of local languages.* India is home to several hundred languages of which 23 are constitutionally recognized. Hence, the choice of language is another concern. English is the default language supported by all mobile handsets. However, it is not the most suitable language for communicating with farmers. SMS in local languages has been tried by some service providers, but the old, beat–up feature phones owned by farmers do not support local language SMS.[1]
- *Top-down generic information.* A review of ICT projects in agricultural extension (Balaji *et al,* 2007) makes the following inferences: (i) localization and customisability of content are still not practiced on a significant scale; (ii) there is a prevalence of top-down approaches with few attempts to reflect the end users' preferences and needs.

Typically, service providers register farmers and once registered, information selected by the service provider is pushed to the farmer's phone, either as SMS or voice message, at pre-defined intervals. Some services allow content to be pulled by querying a database. However, the limitation in both cases – push and pull – is that the information is not tailored to the unique situation being faced by the farmer. Even after obtaining the information, the farmer may still have to fall back on traditional sources such as peer

group or input service providers. Given that provision of agricultural extension services is one of the primary uses of mobile phones in agriculture, it is pertinent to examine whether farmers find the information timely and relevant to their needs. As one review paper puts it:

> Unfortunately, most mobile-based agro-advisory initiatives of the public sector provide routine information with limited refinement and validation.
>
> (Saravanan and Bhattacharjee, 2015, p. 40)

The lack of location-specific and farmer-specific content and the need for timely information have also been analysed as shortcomings of the current services.

To overcome some of the challenges mentioned above, especially the lack of localized and customized content, the following sections describe a model that builds upon a comprehensive farmer database to provide farm-specific information. The model is based on ICRISAT's ongoing work to help improve rural livelihoods by scaling up science-led research for development. The approach integrates the use of ICT with other crop/site-specific technological and agronomic interventions to help raise productivity and income of small and marginal farmers in dryland areas. The ICT interventions described below have been earlier piloted in Addakal in Telangana and Ananthpur in Andhra Pradesh states in India. Based on the learnings and feedback, the model is being rolled out in four districts of Karnataka state in India.

3.4 Targeted Information Delivery

A pilot project 'Krishi Gyan Sagar' (KGS), designed[2] to deliver farm-specific information, is being piloted in four districts[3] (Raichur, Bijapur, Tumkur and Chikmagaluru) of Karnataka state in India. These districts rely primarily on rainfed agriculture and are prone to droughts. Climate trend analysis shows increasing temperature trends at all the sites along with decreasing moisture availability indices resulting in increased aridity (ICRISAT 2015).

KGS is a database-driven software which has two parts: (i) an Android app which can be accessed via an Android-based tablet; and (ii) a web app which can be accessed on a desktop/

laptop (*see* Fig. 3.1). The Android app, which can be accessed on a tablet, is designed as a mobile data collection tool as well as an information dissemination tool. Farmer facilitators use this app to register farmers and collect farm-level data using the application. They also provide targeted information to farmers such as: soil-test-based crop-specific fertilizer recommendations and a crop-specific improved package of practices in the local language. This information is tailored for a particular farm based on the specific farm details available in the database. Availability of information is restricted based on the jurisdiction of the logged-in user.

The web application is used by policy makers, government officials, research and development agencies for monitoring and report generation. The primary government users from the Department of Agriculture, Government of Karnataka, are: Director, Joint Director, Additional Director, Agriculture Officer and Assistant Agriculture Officer, while other users are scientists, scientific officers and research technicians. Users can generate reports based on the data captured by the farmer facilitators. Both the

Android app and the web app connect to the same database server.

The hardware used in the pilot is Samsung Galaxy Tab 2. However, any tablet with similar specifications – 7-inch touchscreen, 3G and Wi-Fi connectivity, GPS, Bluetooth, voice calling facility, primary and secondary camera with good resolution, minimum 1 GB RAM with expandable memory and running on Android OS – is suitable. Since this tablet will be used outdoors, ruggedness of the gadget is the most preferable feature. The app, available in English as well as the local language, has modules as described in the following sections.

3.4.1 Farmer registration and details modules

A comprehensive farmer database is the foundation of any effective agricultural extension system. This module (*see* Fig. 3.2) captures farmers' details in the database. The fields include basic demographic information such as name,

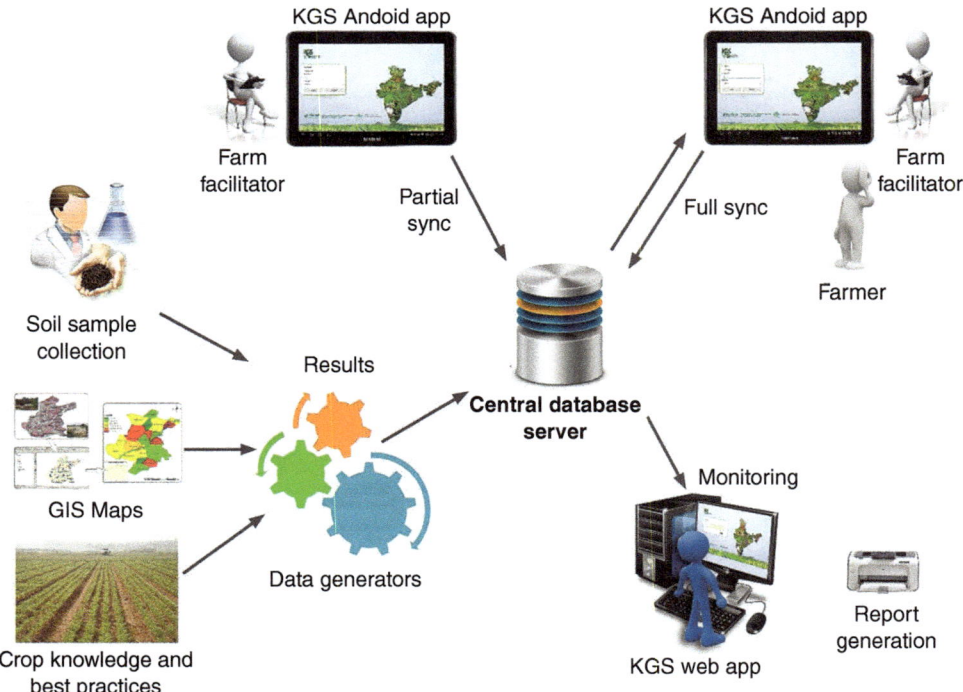

Fig. 3.1. Schematic of the KGS system.

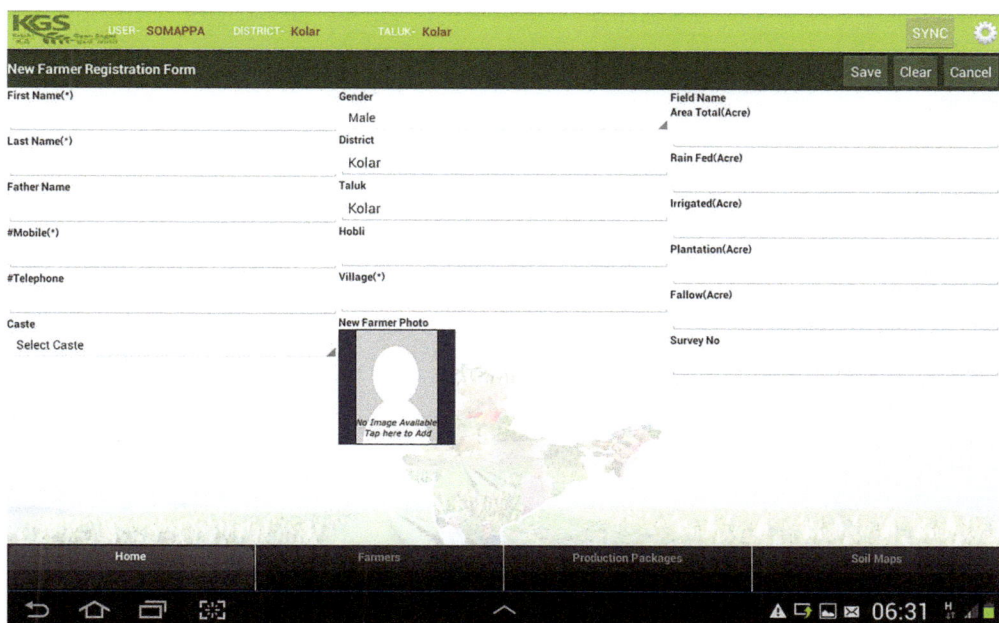

Fig. 3.2. Screen shot of farmer registration module in the KGS app.

location, contact details, and photograph. A unique identity number is generated for each farmer. The basic information is captured once and can be edited any time. Previously this information used to be collected on paper forms when farmers came to local extension centres for taking inputs. So far around 10,000 farmers have been registered. Once a farmer is registered and the basic details captured, detailed farm information such as the number of parcels of land, location of land, crops grown in each season, area of rainfed land, area of irrigated land, area left fallow, source of seeds, government subsidy taken, and fertilizer usage is captured. Geo-referenced location data for each parcel of land is automatically captured.

The location information of each farmer is linked with geo-referenced soil fertility data, which is used for providing site-specific fertilizer recommendations.

3.4.2 Soil test-based fertilizer recommendation

The KGS pilot relies on the soil fertility atlas (Wani *et al.*, 2011) (*see* Figs 3.3 and 3.4) which

has been created for the state of Karnataka. These data are used in the KGS app in two forms: (i) district-wise soil fertility maps including status of organic carbon, phosphorous, potassium, sulfur, boron, and zinc; and (ii) site-specific fertilizer recommendation. Based on the user's district, the appropriate soil maps will be displayed in the app. KGS is backed by geo-referenced soil fertility data and individual farmer's location information recorded at the time of registration. The queried data is processed on the basis of location, farm area and crop-specific nutrient requirements to provide customized fertilizer recommendation.

3.4.3 Package of agronomic practices and pest control module

This module (*see* Fig. 3.5) provides updated information about good agricultural practices with respect to each crop grown in the region. It contains information about soil and climate requirements, land preparation, available cultivars, seed treatment, sowing/planting, fertilizer and water management, plant protection practices, harvesting and post-harvest practices etc, based on

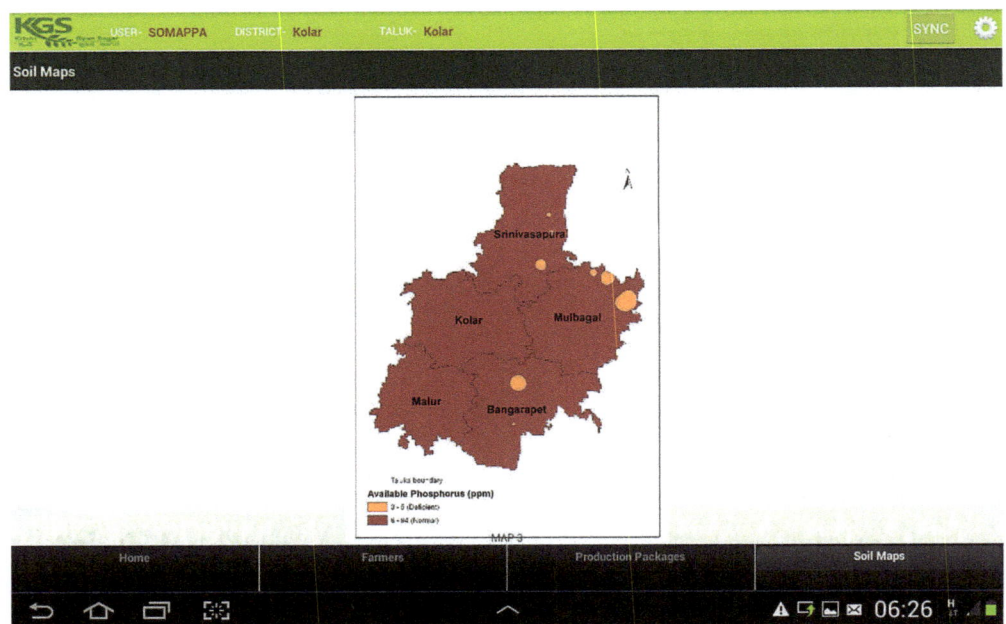

Fig. 3.3. Soil fertility map displayed in the KGS app.

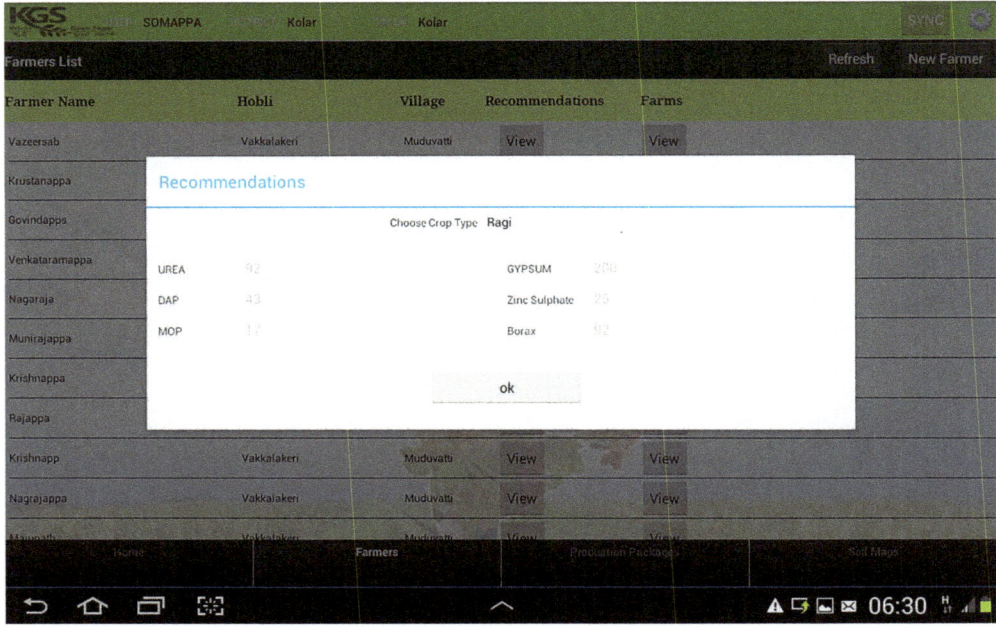

Fig. 3.4. Crop- and location-specific fertilizer recommendation for each farmer.

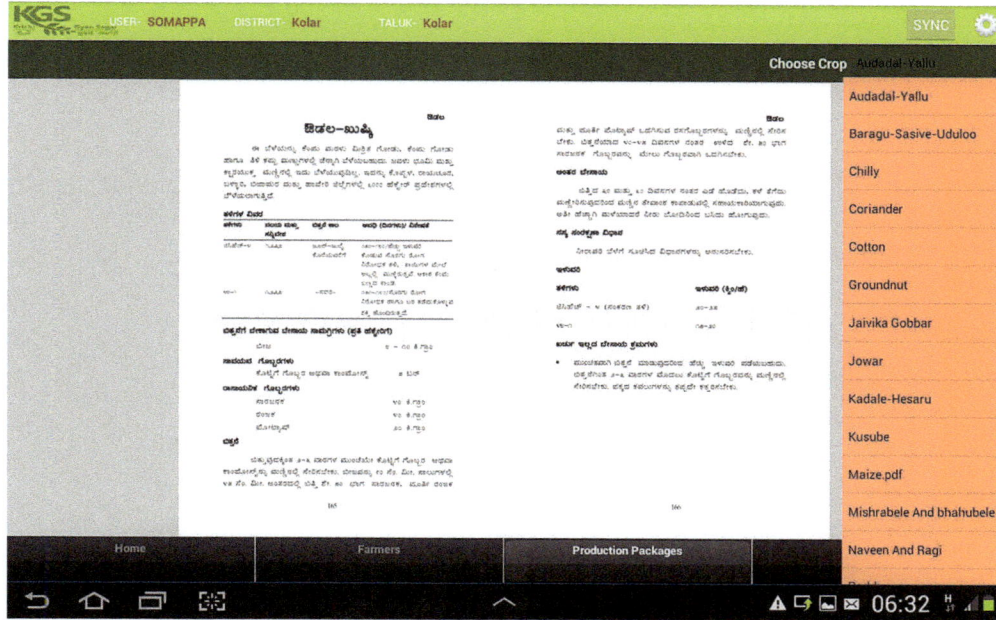

the recommendations of the state agricultural universities. Crop-specific pest control information is also available in this module. This information is available in the local language and supported with adequate illustrations and photos so that the farm facilitators can easily understand it, and in turn convey it to farmers.

3.4.4 Additional modules

There are additional modules for the farmer facilitator:

- *Field visit module* – this module is for recording the field visits undertaken by the farmer facilitator. Details captured are date, location, purpose, observations and comments. The module also has the facility for uploading field visit pictures.
- *Training module* – all training provided is recorded in this module. Details captured are date, location, topic of training, level and type of training, and number of participants (disaggregated by gender). There is also the facility for uploading training pictures.
- *Crop cutting estimate module* (for yield estimate).

The above information resides in the database and is updated periodically by the farm facilitator. For example, at the start of the cropping season, the crops the farmer intends to plant are recorded; type and amount of fertilizer used is recorded; any abnormality during the growing season is recorded at the time; and the yield is recorded at the time of harvest.

Fig. 3.6 shows the process of data collection and synchronization of the KGS app. When the farmer facilitator visits an area the Android app synchronizes with the database and downloads the complete information of farmers in that area, soil fertility maps and package of practices for crops grown in that area. The farmer facilitator can then update farm information, if required, information about his/her field visits, training conducted, results of crop cutting

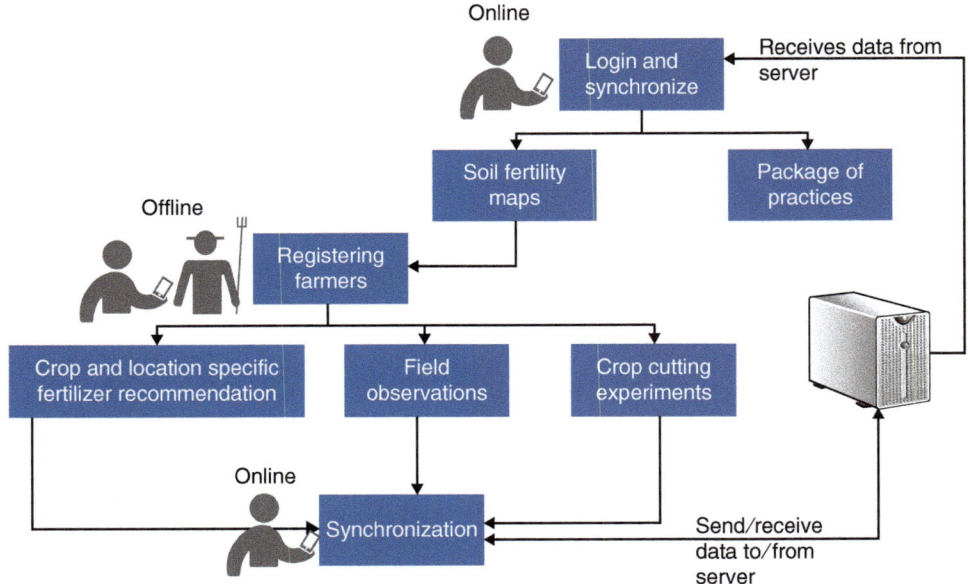

Fig. 3.6.   Flow diagram of field operation of KGS app.

experiments, etc. She/he can also extract any information required by the farmer by querying the database; conveying information about a package of practices for specific crops; and supplying farm-specific fertilizer recommendations based on soil health records.

To overcome connectivity problems, common across rural areas of the country, all data recording processes are offline by which the data is stored in the device. When connectivity is available a partial synchronization process uploads all updated data stored in the tablet to the server.

Karnataka while the ICT interventions are being piloted in 4 districts. While it is too early to draw any definite conclusions, comparative crop yield data for farmers who are part of the project show encouraging results. Groundnut cultivars produced 35–40% higher yields, whereas pigeon pea cultivars produced 40–45% higher yields under deficit rainfall situation compared to farmers' cultivars. Castor cultivars produced 25–37% higher yields while finger millet productivity was 63% higher (ICRISAT 2015). Analysing the specific impact of ICT interventions on farm productivity and incomes is in progress.

3.5 Conclusion

The project started in the 2013–14 cropping season. The ICT intervention was designed to supplement the ground-level interventions in the districts to improve farmers' livelihoods through improved soil and water management, integrated nutrient management, integrated pest management, introduction of improved cultivars, improved agronomic practices and linking to markets. The agricultural interventions are being implemented in all 30 districts of

Notes

[1] Author's observations during a field visit to a SMS-based crop sowing advisory project in Kurnool district of Andhra Pradesh, India.
[2] See: https://www.cabi.org/cabebooks/ebook/20163299419 (accessed January 2018)
[3] A district is the basic administrative unit in India. State are divided into districts which have further sub-divisions known by various names, such as *block*, *mandal*, *taluk* and *thesil* in different states.

References

Balaji, V., Meera, S.N. and Dixit, S. (2007) ICT-enabled knowledge sharing in support of extension: addressing the agrarian challenges of the developing world threatened by climate change, with a case study from India. *SAT eJournal* 4(1), 1–18.

Banerjee, S. (2013) Mobile telephony in agriculture: unlocking knowledge capital of the farmers. In: Sylvester, G. (ed.) *Information and Communication Technologies for Sustainable Agriculture. Indicators from the Asia and the Pacific, Part II – Technologies for Agricultural Information Sharing.* FAO Regional Office for the Asia and the Pacific, Bangkok, Thailand.

Cole, S.A. and Fernando, A.N. (2012) The value of advice: evidence from mobile phone-based agricultural extension. *Working Paper 13-047*, Harvard Business School, Harvard University, Cambridge, Massachusetts, USA.

Dhaka, B.L. and Chayal, K. (2010) Farmers' experience with ICTs on transfer of technology in a changing agri-rural environment. *Indian Research Journal of Extension Education* 10(3), 114–18.

ICRISAT (2015) *Annual Report 2014–2015 Bhoochetana Plus: Improving Rural Livelihoods through Innovative Scaling-up of Science-led Participatory Research for Development.* ICRISAT, Patancheru, India.

Kameswari, V.L.V., Kishore, D. and Gupta, V. (2011) ICTs for agricultural extension: a study in the Indian Himalayan region. *Electronic Journal of Information Systems in Developing Countries* 48(3), 1–12.

Mittal, S. (2016) The role of mobile phone-enabled climate information services in gender-inclusive agriculture. *Gender, Technology and Development* 20(2), 1–18.

Mukherjee, A., and Maity, A. (2015) Public–private partnership for convergence of extension services in Indian agriculture. *Current Science* 109(9), 1557–1563.

Saravanan, R. and Bhattacharjee, S. (2015) Mobile phone applications for agricultural extension in India. Available at: http://www.e-agriculture.org/content/mobile-phone-applications-agricultural-extension-india (accessed 21 September 2016).

TRAI (2015) *Telecom Regulatory Authority of India Annual Report 2014–15.* TRAI, New Delhi, India.

Vodafone (2016) Towards a more equal world: the mobile Internet revolution. *Policy Paper Series Number 16.* Available at: http://www.vodafone.com/content/dam/vodafone-images/public-policy/inequality/Vodafone-equal-world-small%20farmers.pdf (accessed 10 August 2016).

Wani, S.P., Sahrawat, K.L., Sarvesh, K.V., Baburao, M. and Krishnappa, K. (eds) (2011) *Soil Fertility Atlas for Karnataka, India.* International Crops Research Institute for the Semi-Arid Tropics (ICRISAT), Patancheru, India, 312 pp.

World Bank (2016) *World Development Report 2016: Digital Dividends.* The World Bank, Washington, DC.

4 mNutrition: Experiences and Lessons Learned in Content Development

Charlotte Day*

CAB International, Wallingford, UK

4.1 Introduction

Mobile devices have contributed to the most innovative changes in peoples' lives in the developing world over recent years. From connecting people through traditional voice calls, to offering safe, flexible banking options, the mobile revolution is not slowing down.

Coupled with runaway growth in mobile ownership, connectivity and functionality, the majority of people in the developing world (70%) are engaged in agricultural activity for their livelihoods through their own production or employment. Access to sufficient agricultural information is a challenge to many farming communities in developing countries because of poor transport and communication infrastructure and a reliance on overstretched traditional agricultural extension services (FAO, 2014).

Bridging the agricultural knowledge gap through a fast-growing industry such as mobile communication holds impressive promise. A report commissioned by Vodafone identified 12 mobile-enabled solutions for food and agriculture grouped into improvements in financial services, provision of information, data visibility for supply-chain efficiency, and enabling access to markets (Vodafone, 2011). Implementing these opportunities would create an estimated

US$138 billion in agricultural income across 26 countries in 2020, 11 percent more than the previously projected value. Specifically related to these mobile opportunities, the report highlights that 'opportunities for mobile technology lie mainly in supporting smallholder farmers in the primary production and marketing processes . . .' (Vodafone, 2011, p. 10).

Identified as good candidates for delivery of behaviour interventions (WHO, 2011), a few value-added service (VAS) offerings are attempting to test the feasibility of mobile communications for behaviour change in agricultural practices (GSMA, 2014), especially related to the impact this has on household nutrition (GSMA, 2015).

The GSMA's mNutrition Initiative, funded by DfID, sought to bridge the business interests of mobile operators with this development aim, by working with mobile operators and other partners in 12 countries in Africa (Ghana, Nigeria, Malawi, Tanzania, Mozambique, Uganda, Kenya and Zambia) and South Asia (Sri Lanka, Bangladesh, Pakistan and Myanmar) between June 2014 and May 2017 to provide commercially sustainable agriculture and health VAS to the rural poor. This chapter focuses on the content-related experiences and lessons of the mAgri component to mNutrition, as reported by content partners of the initiative.

---

* E-mail: C.Day@cabi.org

4.2 Outline of mNutrition

The GSMA, in partnership with the UK government's Department for International Development (DfID), launched the mNutrition Initiative in 2014 to develop and scale-up the delivery of nutrition- and agriculture-related services using mobile-phone-based platforms until 2017. As set out in the request for proposals in the initial project documentation, the aim of the initiative was:

> Improved nutrition for the poor as a result of behaviour change promoted by accessible mobile-based services delivered at scale through sustainable business models.

and the GSMA's project theory of change specified that it was expected to:

> reach at least three million people across ten sub-Saharan African and four Asian countries. The change in nutritional status of the target population is expected to occur due to the following – support change as it relates to nutrition, registration of target populations, cultivation, and consumption of crops with high nutritional value(s) and; timely and efficient data surveillance of key nutrition indicators. This will be complemented by improved access to food as a result of improved agricultural production and income.

The GSMA delivered this through leveraging expertise and capacity from two of its existing development initiatives under Mobile for Development: mFarmer (mAgri) and mHealth. As specified in the request for proposals, the approach was to:

> bring together mobile network operators, the wider mobile industry and the development community to develop, test and roll-out commercial mobile services for the underserved people in emerging markets.

This includes developing a number of 'streams' within the initiative, and partnering with selected organizations to help deliver on these. Streams include: service development user experience (UX) for the mobile service design; monitoring and evaluation (M&E), which could provide insights into usage and identify opportunities for improvements to service design and content; consumer research, which could be fed into content development; business intelligence (BI) on the service usage and acceptability;

nutrition-related content development; and fund management (for mAgri only). The project plan demonstrated that there would be a limited phased implementation, delivering 'waves' of countries at a time. The implementing countries were divided into waves as laid out in Table 4.1.

Leading the content creation stream was a consortium – the Global Content Partners (GCP) – consisting of CAB International (CABI), the British Medical Journal (BMJ), the Global Alliance for Improved Nutrition (GAIN), Oxfam GB and the International Livestock Research Institute (ILRI). The role of the GCP was to:

- Develop a general framework for the development of nutrition content, from which a local version can be developed for use in each country.
- Establish a network of authoritative local organizations to guide and support the implementation of health and/or agricultural information services.
- Carry out a landscape analysis of current state of nutritional needs, content availability, stakeholders, pricing and regulation for nutrition and evidence-based intervention practices in each of the project countries and identify key factors for sustainable content services beyond the project.
- Contract and provide technical assistance to a local content organization to partner with mobile operators to either scale up existing or develop, launch and market new mNutrition content services.

Table 4.1. mNutrition countries by wave of implementation.

Wave 1 – mAgri	Wave 2 – mAgri	Wave 1 – mHealth	Wave 2 – mHealth
Bangladesh	Pakistan	Ghana	Uganda
Sri Lanka	Myanmar	Nigeria	Zambia
Malawi		Malawi	Mozambique
Ghana		Tanzania	Kenya
			*Rwanda
			**Côte D'Ivoire

*Rwanda did not remain a full implementing country of the mNutrition Initiative, however a refined package of content activities was delivered.
**Cote D'Ivoire was withdrawn from the Initiative as an implementing country.

4.3 The mNutrition Content Development Process

Following early discussions with project partners, it was highlighted that one of the key factors that had been holding back the development of agriculture, health and nutrition mobile content services was the lack of trusted partners in-country to provide high-quality content that met the needs of the local population, service providers and key stakeholders, such as government authorities. At the same time, content developed centrally by trusted international organizations (e.g. the World Health Organization), while technically accurate, lacked the insights into local needs, motivations and barriers to change, which are required if messages are to have a chance to bring about positive changes in behaviour. Therefore, it was decided that content should be produced by Local Content Partners (LCPs) who could create locally valid content, and that this content would be relevant to nutritional gaps identified in the national landscape studies produced by the GCP within the project.

Based on this, the GCP set about to identify local, motivated content organizations with potential to be supported by the GCP with:

- A robust process for sourcing, developing and quality assuring content for mNutrition, and training on this.
- Assistance in identifying and adhering to national priorities and getting messages validated through introductions to key stakeholders and landscape analysis.
- Support to ensure alignment with national priorities and engagement with government authorities and other key stakeholders, which would also increase the chances of mNutrition messages achieving impact as they could complement and amplify any existing nutrition messaging through other channels ongoing mentorship.

Prioritization of nutritional need gaps, the national nutrition and/or agricultural policy framework and high-level partnerships with organizations, such as the Scaling Up Nutrition (SUN) national bodies and government ministries, would come from the GCP, whilst insights into local needs, motivations and barriers to behaviour change would come from the LCP,

supplemented by insights from the consumer research work stream (and subsequently from the M&E stream).

Within the scope of the content stream is the creation of an open-access content repository, where all content produced under the mNutrition Initiative could be stored and available for further users to access and repurpose. A schematic of how the content process intended to flow is provided in Fig. 4.1.

4.3.1 Assumptions

The premise was that this combination of a local, motivated team with some editorial experience, and the GCP's robust editorial process with appropriate training and supervision from GCP, should lead to high-quality content. GCP's assumptions underpinning the decisions made during the design of the content process were as follows:

- Critically, if the content services are to stand any chance of being sustainable beyond the life of the mNutrition programme, the emphasis must be on getting LCPs to follow this editorial process and take ownership of the local content they produce (along with key stakeholders). Through self-assessment (and GCP oversight and steer) LCPs would ensure adherence to the prescribed quality assurance (QA) procedure.
- LCPs would be able to form a relationship, with support from GCP and GSMA, with the service providers for insights into users, access to target audiences for end user testing, and feedback loops from service usage. This would also allow both LCPs and service providers to prioritize content development.
- Quality Control (QC) must be objective and refer to process, so instead of saying 'this message doesn't seem relevant', ask for evidence that the message has drawn on customer insights and/or has been tested with users. Limited resources mean that QC must be carried out on a sample of LCP content with a greater frequency at the beginning. This should enable GCP to identify any patterns of poor practice early on and

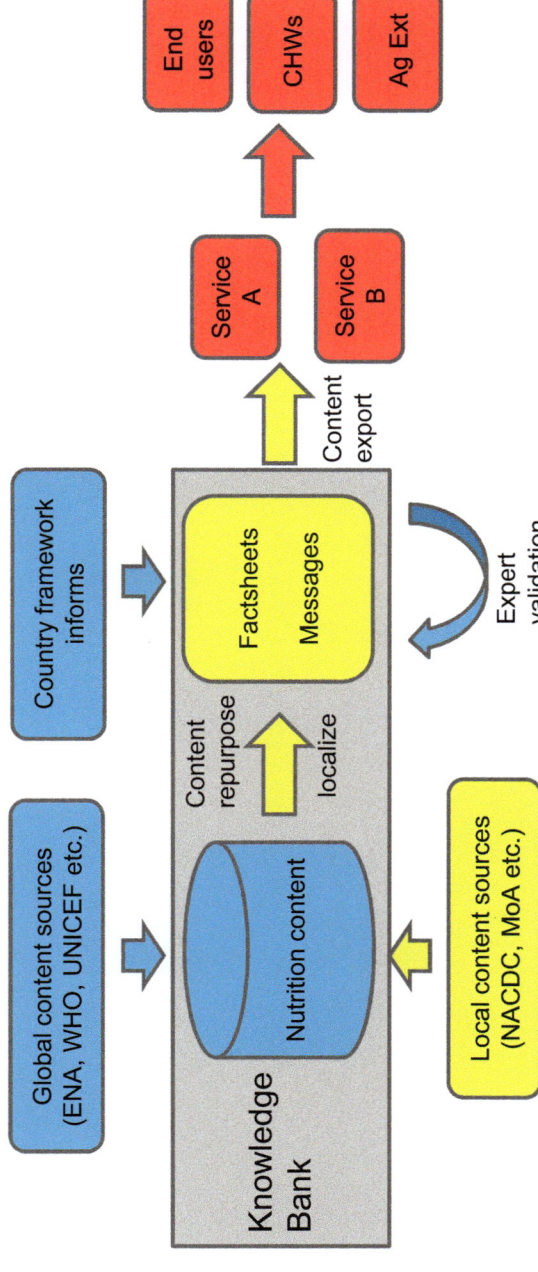

Content Process Flow

Knowledge Bank

Global content sources (ENA, WHO, UNICEF etc.)

Country framework informs

Nutrition content

Content repurpose localize

Factsheets Messages

Local content sources (NACDC, MoA etc.)

Expert validation

Content export

Service A

Service B

End users

CHWs

Ag Ext

GCP (consortium)

Local content provider

Service providers

- Global and local validated source content will be uploaded to a central database – the nutrition Knowledge Bank.
- Local Content Partners will create health and agri focussed nutrition factsheets by country.
- Factsheets will be validated in country by relevant stakeholders (e.g. government agencies).
- Local Content Adapters will create sets of key nutrition based mobile messages.
- Mobile messages will be validated in country by relevant stakeholders (e.g. government agencies).
- All content will be centrally stored on the nutrition Knowledge Bank for use by multiple mobile services.

Fig. 4.1. Schematic of content production process and use by mobile operators.

get the LCP to take corrective action across all content.

- LCPs understand the gap analysis and can source content relevant to it.
- In anticipation of a wide range of capabilities and experience and without the opportunity to pilot different approaches, editorial process and training materials can't be too prescriptive, and must allow some flexibility while at the same time ensuring that a consistent approach is followed.
- Content would be developed per priorities agreed with key national stakeholders (and captured in a country content framework) and tailored to address needs, motivations and barriers faced by target audience, rather than being tailored to a specific service (therefore, the relationship with these key partners is a prerequisite so that the content could be used by a number of different service providers and adapted by them (stylized) to meet service-specific requirements).
- Central repository of all factsheets and messages would maximize potential for content to be reused and repurposed both within the mNutrition programme and beyond.

At the end of the initiative's implementation, in addition to seeing mNutrition messages having the desired impact, a goal for the GCP was to see local content organizations that are now trusted to produce high quality content and can be commissioned by service providers to do so without the need for support from GCP.

Whilst there are many steps within the content production process (provided in a generic format in Fig. 4.2), four highlighted areas are discussed further in this chapter: quality, localization, partnerships and sustainability. These were selected due to their cross-cutting nature and importance for widespread applicability to other agricultural content creation projects and contexts.

4.4 Lessons Learned

4.4.1 Quality processes and criteria

Quality, whilst arguably very subjective by nature, is a key component of content creation. This is in order not only to ensure a benchmark to which content should adhere for reasons of

Content Production Process

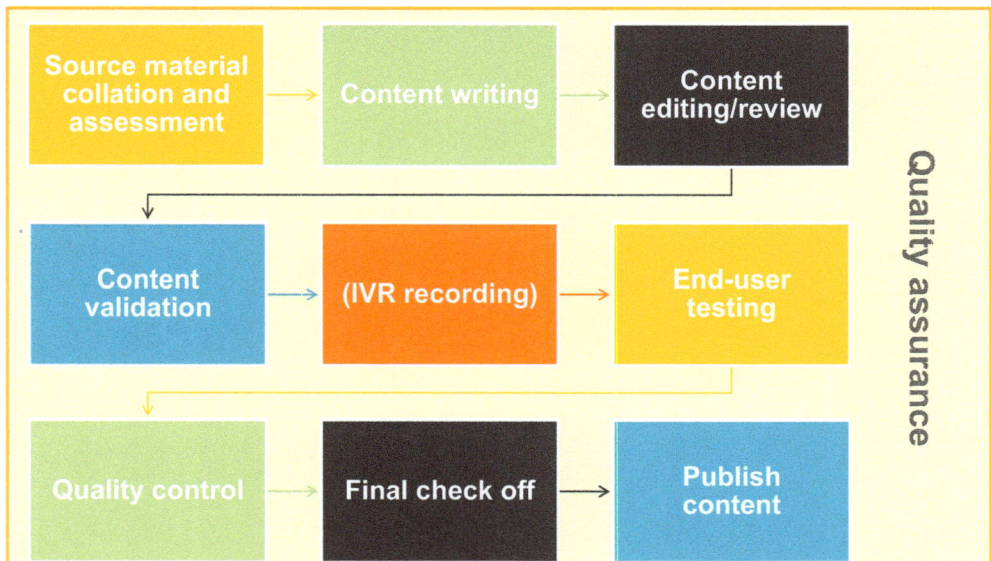

Fig. 4.2. Simplified content creation process flow.

effectiveness, but also to instil confidence amongst target beneficiaries, wider project stakeholders and content teams themselves. This has the added further benefit of creating a stronger sense of trust between content teams and intermediaries – such as mobile network operators and service providers – that an ongoing relationship for new or refreshed content will be derived and created from a reliable source.

The GCP created a series of quality assurance methods that were all checked as part of a quality control (QC) process conducted by ILRI on behalf of the GCP. The methodologies and their objectives are given in Table 4.2.

There are four key quality principles identified of which each compromises a set of quality criteria. These are presented in Table 4.3, along with a description of which quality assurance methodology should be employed to ensure content adheres to the given criteria.

Implementing partners agreed that content created by in-country partners was the most appropriate content development model, as LCPs offer the best change of sustainable relationships at the country level so that beyond the project lifetime, the content process and relationships in place could continue. Based on the assumptions given above, the implementation of

Table 4.2. QA methodologies used to assess mNutrition content.

QA methodology	Description
Country Prioritization	Identify key crops and livestock for content coverage, and appropriate priority interventions and recommendations
QA Process	Check content is accurate, safe, effective, referenced, reliable and grammatically correct
End-user testing (pre-release)	Confirm content is user friendly
Validation	Confirm content meets national requirements
Quality Control	Check QA process has been followed
End-user testing (post-release)	Confirm content is understandable and remains on message

Table 4.3. Quality principles and QA assessment.

Principle	Criteria	Description	QA methodology
Reliable	Validated	Approved for use by national/local experts	Validation
	Trustworthy	From a reliable and trusted source	QA process
Relevant	Specific	Crop- and context-specific	QA process
	Timely	Time of day, time of crop cycle, incident alert	QA process End user testing (post)
	Localized	Correct geography, production system, varieties, etc.	Country Prioritization QA process Validation End user testing (pre and post)
Clear	Language	Non-scientific, gender appropriate and in correct local language	End-user testing (pre and post)
	Tone	In a trustworthy and accessible 'voice'	End-user testing (pre and post)
	Importance	Why the action will be of benefit	Country Prioritization Validation End-user testing (pre and post)
Practical	Actionable	Provides suggestions and solutions that end user can act up	Country Prioritization End-user testing (pre and post)
	Realistic	Is appropriate in the given context, and with the end-users' given resources	End-user testing (pre and post) Validation

the content process and indeed corresponding quality assurance relied on the pre-existing capacity, willingness to learn and adapt, and process ownership of the LCP. Given the stringent timeframes for content production and publication within this initiative, building sufficient capacity and ownership at the local level proved continually challenging. Extra 'hands on' support was required from the GCP to ensure timelines and quality standards were both met, however, one or the other was regularly compromized.

Recommendation: build sufficient time buffers for content providers to complete/iterate the processes requested of them: content creation, review, approval, QC, validation, translation, end-user testing etc., before content is required for a launch/addition to service.

The introduction of the QC gateway came about approximately halfway through the majority of LCPs' contracts, when an enhanced QC process, based on criteria covered in initial LCP training, became a 'gateway' for the release of content in which content not meeting the quality criteria was returned to LCPs for further editing before it was published. This was because ILRI's QC reviews and GSMA observed that the same quality problems were recurring even when they had been identified and corrected in previous reviews, and when content had also been validated by national stakeholders.

Since this was implemented, the overall quality of content produced by LCPs did improve. However given the substantial amount of support required from GCP to LCP, and the multiple steps involved in content production, the subsequent impact of this step was on content completion timeframes, which were extended, sometimes to beyond the LCPs' contract period.

More critically, the major implication of this is the reduced opportunity for LCPs to take ownership over content (identified as a key requirement of content creation) and instead increasing a reliance on external support and input from GCP. The whole content process was designed to provide a decreasing reliance on such external support from GCPs as LCPs capacities developed, thereby increasing the likelihood of more sustainable relationships between LCPs and MNOs and VAS providers in country. This outcome was not realized in every context due to the 'quality' limitations of some LCPs, but also

due to this external reliance as introduced by the gateway.

Recommendation: define who content needs to be 'good' for, by employing the MoSCoW[1] method, and provide this upfront to all parties. In addition, guidelines on roles, responsibilities, expectations, process and criteria should be made upfront to content partners.

Recommendation: provide sufficiently clear support tools to better ensure adherence to the defined process.

Recommendation: ensure that there is capacity to conduct QC reviews in the language in which users will receive the content to avoid unnecessary and limited rendering of local language script.

4.4.2 Localization

In September 2014, GSMA published an industry intelligence report entitled *Local World – Content for the Next Wave of Growth* (GSMA, 2014) in which local content is identified as:

> the key driver in creating a step change in the usage and engagement of the mobile internet and mobile-enabled services, particularly for mid and low income consumers in emerging markets.
>
> (GSMA, 2014, p. 8)

It further suggests that driving uptake in this demographic is particularly important as it was seen to be the primary growth segment in the industry in the following five years. The report also defines, discusses and differentiates between types of local content: available in local language; locally relevant information; and locally created. Although content may be available in a local language it may not be particularly relevant to local audiences. Local creation is deemed to be the best way forward, but it is not easy, particularly in markets where there is a lack of data on consumer insights. If expert-generated content is to be used, it must engage with users in a 'user-centred design process' to ensure the content is really something users find interesting and relevant.

Recommendation: end-user testing should be established as a crucial component of the content development process. Conduct end-user testing pre-release of content, so that feedback can be incorporated before content is rolled out through the service. Creating content based specifically on user-demand is a costly and time-consuming endeavour and requires field research into specific local practices, interests and details, which are not well documented in the existing body of literature. Ensure that sufficient time is allocated to this activity – including incorporating learnings – as part of the content development process.

In some cases, where content was written in a local language by the LCP, the translation to English purely for QC purposes was a largely redundant activity, particularly when recommended edits involved nuances of language, and corrections to the local language became complex for the LCP to undertake.

Recommendation: where possible, divert efforts from creating 'base' material in English (if this can be sourced or repurposed from elsewhere) and place emphasis on measures to localize.

4.4.3 Partnerships

Partner networks required for content development are complex and from a range of sectors. Regardless of the number of implementing countries within the mNutrition initiative, there are many key relationships that need to be in place, and which are strongest when there are pre-existing relationships to build on.

For the GCPs and LCPs involved in each country, typical additional relationships are required with: MNOs and/or service providers; the appropriate government body for content validation and sign-off; local content experts such as civil society groups, UN agencies, universities and other project partners including GSMA, UX and M&E teams.

Through the experiences of implementing the mNutrition initiative in 12 countries, the key characteristics of good partnerships are listed as recommendations below.

Recommendation: open communication: from project design through implementation to wrap-up, the complexity of stakeholder relationships requires clear, consistent and open communication channels to achieve effective delivery throughout. Whilst this may sound obvious, good communication is often taken for granted and therefore due consideration is required during the design and planning stages to lay out a communication framework, which itself is well communicated and understood by all.

Recommendation: collaborative planning: in line with open communication, it is imperative that key stakeholders are engaged in the planning stages in order to outline expectations of all parties from the outset, and to secure investment from a range of project stakeholders throughout. This may therefore require more time and effort up-front, but will undoubtedly ensure implementation is more efficient and effective. This has the added benefit of aligning key project milestones with those of others, building a stronger 'campaign' approach to message dissemination.

Recommendation: flexibility to different and changing priorities and adherence to agreed processes: recognition that, especially when subcontracts are in place, changes in programme design must acknowledge the principles, objectives and scope of existing agreements, so that a full re-working of these contracts are avoided, thereby saving time and reducing likelihood for friction at multiple levels of the programme.

4.4.5 Sustainability

Given the complex nature of this cross-sector project, and the experiences of the content partners involved, it is apparent that in order to achieve sustainable content creation, update and dissemination systems, key sustainability objectives need to be included from the beginning. These evolve around more capacity building, local ownership and local relationship building.

Recommendation: supporting capacity of LCPs to continue to follow a rigorous, quality-driven content process. Embed quality assurance and process checks within organizational ways of working. Allow LCPs to take ownership of content quality to ensure 'buy-in' to the benefits of high-quality content and the reputation which supports that.

Recommendation: support LCPs to facilitate and nurture good relationships with key in-country partners, such as government, service providers and MNOs.

Recommendation: showcase the achievements of the LCPs and the opportunities available to them for repurposing content. This also contributes to their reputation as a quality content provider, which itself ensures the LCP is more likely and better prepared to receive future content-related work.

4.5 Conclusion

The mNutrition Initiative, delivered by GSMA and supported by a CABI-led consortium of global partners to deliver the content stream, provided many significant lessons on local content development practises. These can be categorized under four main themes: quality, localization, partnerships and sustainability.

The common driver of each theme is process ownership by the LCP. Whilst decentralising the entire content production process has its challenges and is likely to result in a less coherent delivery approach at the global level, it ensures much smoother delivery of the priority actions at the local level, which in turn better achieves localization, quality, partnerships and sustainability.

Note

[1] MoSCoW method refers to a prioritization technique to ensure all stakeholders are familiar with the importance each requirement or deliverable is given. These are categorized as: Must have, Should have, Could have and Won't have.

References

FAO (2014) Communication for rural development: guidelines for planning and project formulation. Available at: http://www.fao.org/3/a-i4222e.pdf (accessed 16 February 2017).

GSMA (2014) Local world – content for the next wave of growth. Available at: https://www.gsmaintelligence.com/research/?file=5fdd9e71266463b59f7e21a08843d0f7&download (accessed 15 February 2017).

GSMA (2015) Mobile for development mHealth: the use of mobile to drive improved nutrition outcomes: successes and best practices from the mHealth industry. Available at: http://www.gsma.com/mobilefordevelopment/wp-content/uploads/2015/04/M4D-mHealth-improved-nutrition_R1_web.pdf (accessed 15 February 2017).

Vodafone (2011) Connected agriculture: the role of mobile in driving efficiency and sustainability in the food and agriculture value chain. Available at: https://www.accenture.com/mu-en/_acnmedia/Accenture/next-gen/reassembling-industry/pdf/Accenture-Connected-Agriculture.pdf (accessed 15 February 2017).

WHO (2011) mHealth: new horizons for health through mobile technologies. Available at: http://www.who.int/goe/publications/goe_mhealth_web.pdf (accessed 15 February 2017).

5 Introducing a Technology Stewardship Model to Encourage ICT Adoption in Agricultural Communities of Practice: Reflections on a Canada/Sri Lanka Partnership Project

Gordon A. Gow*,†

University of Alberta, Edmonton, Canada

5.1 Introduction

Information and communication technologies (ICTs) have long been regarded as forces for positive change in agriculture and rural development despite a track record of modest success with many initiatives (Duncombe, 2012). ICTs are considered, particularly among frontline development practitioners, as important tools for mobilizing knowledge because they can lower transaction costs associated with information seeking and because they can introduce new social practices for improving farmer education and training through the use of various forms of digital media (Farm Radio International, 2011; O'Donnell, 2011). Low-cost technologies such as the mobile phone have been the subject of intense focus within the ICT for Development (ICT4D) community as holding considerable promise for knowledge mobilization in the agriculture sector (Qiang *et al.*, 2011).

However, difficulties in creating and sustaining capacity, as well as interest, within communities to adopt and use low-cost ICTs for knowledge mobilization activities has prompted leading scholars in the ICT4D community to consider inclusive innovation models that emphasize direct participation of the community in establishing information services and related social practices, such as market access information and knowledge networks that can foster iterative learning and long-term capacity building within those communities (Foster and Heeks, 2013, 2014; Heeks *et al.*, 2014).

A combination of low-cost technologies with various open-source software platforms for text messaging, crowdmapping, and interactive voice response systems has created a new opportunity to take up and examine more closely the possibilities for articulating an inclusive innovation approach, using a technology stewardship

* E-mail: ggow@ualberta.ca

† Additional contributions from: Helen Hambly, Associate Professor, University of Guelph; Udith K. Jayasinghe-Mudale, Professor, Wayamba University of Sri Lanka; Uvasara Dissanayeke, Sr Lecturer, University of Peradeniya; Rob McMahon, Assistant Professor, University of Alberta; Nuwan Waidyanatha, Sahana Software Foundation; Chandana Jayathilake, PhD student, Wayamba University of Sri Lanka; Sean McDonald, CEO, FrontlineSMS; Faria Rashi, graduate student, University of Guelph; and Ken Lee, graduate student, University of Alberta.

© CAB International 2018. *Digital Technologies for Agricultural and Rural Development in the Global South* (ed. R. Duncombe)

model being developed and deployed on a trial basis in Sri Lanka through a partnership with Canadian and Sri Lankan researchers and practitioners.

Technology stewardship is an approach adapted from the 'Communities of Practice' literature, for training and supporting individuals and teams who engage their communities to encourage and support innovative practices with low-cost, widely available digital technologies (Wenger *et al.*, 2009, Waidyanatha *et al.*, 2015).

This chapter begins with a discussion of the theoretical framework informing the project, describes the stewardship approach as it has evolved so far, and then presents preliminary results of an ongoing action research project in Sri Lanka where we have conducted a series of pilot studies in technology stewardship with partner organizations. The chapter concludes by outlining plans for a 'Joint Education and Training Initiative' for technology stewardship launched in September 2016.

5.2 Development and ICTs

As Kleine (2013) has noted, the 'intellectual space for ICT4D' represents a wide terrain and even the term 'development' itself is contested among scholars and practitioners. This project takes a perspective shared by a school of thought that aligns with Sen's capabilities approach, which in simplified form argues that development should be regarded as:

> a process of expanding the real freedoms that people enjoy . . . [the freedom] of people to lead the lives they have reason to value and to enhance the real choices they have.
>
> (Sen, quoted in Kleine, 2013, p. 4)

Kleine draws on this normative statement to describe a set of guiding principles for ICT4D initiatives: (i) development is a dynamic, ongoing, and process-oriented undertaking; (ii) development should focus on enhancing freedom of choice for individuals and, collectively, communities; and (iii) development initiatives should recognize that people themselves can articulate what kind of lives they value and aspire too, while respecting that this will lead to plurality of views:

> any piece of research based on the capabilities approach [therefore] needs to reflect an understanding of development as a process, consider it in a holistic way, and put people at the center, stressing their choices. *The focus on people's choices renders the development process open-ended and pluralistic in its aims.* [emphasis added]
>
> (Kleine, 2013, p. 4)

The emphasis on choice and plurality with the capabilities approach creates a point of tension within the context of a development project, in part, because certain types of outcomes are not necessarily what a funder or research team might have initially had in mind and, moreover, could conceivably result in both positive *and* negative disruptive impacts among the individuals and within the communities affected. This of course has an ethical dimension that practitioners and researchers working within this development paradigm need to consider at all times.

This concern notwithstanding, the main point is that Sen's approach offers a development paradigm that emphasizes the goal of enhancing freedom for individuals through expanded capabilities and the ability to make reasoned choices with respect to those capabilities.

5.2.1 Defining 'ICT' and 'affordable' ICT

The term 'ICT' also needs brief mention to clearly establish a definition and usage convention for the term as taken up within this initiative. ICT stands for 'information and communication technology' and could conceivably encompass 'any technology serving the purpose of gathering, processing, and disseminating information, or supporting the process of communication' (Kleine, 2013, p. 5), including both analog and digital forms of media, tools, techniques, and platforms. However, within the field of ICT4D research, the term more typically refers to digital technologies and, often, some configuration of systems that includes the internet, desktop computers, and mobile phones/tablets.[1]

Specifying the term even further, our initiative has been conceived around the notion of 'affordable ICTs', which is a relative term referring to technologies that are available widely and could be considered affordable for most members of the population. This is an important consideration within the context of the capabilities approach because affordability is a factor that influences the range of choices available to an individual. The term 'affordable' implies terms and conditions of access to a technology in a way that minimizes costs and other barriers (e.g., licensing terms) associated with training and with innovative uses of that technology. Affordability has several considerations in this respect:

1. Affordable access to end-user devices such as mobile phones, tablets, or radio receivers.
2. Affordable access to reliable communication connectivity and services (e.g., commercial cellular or WiFi).
3. Affordable access to user-friendly software and software-based platforms for generating and distributing content and services.
4. Affordable licensing terms to enable continuous integration and customization of devices, software, and software-based services responsive to user needs and ambitions.

Of course, affordability is a measure relative to the population in question and may be an empirical matter that can be established through various metrics. Moreover, affordability for both *producers and consumers* of communications services operating over a configuration of technologies is an important consideration. Following the principles of the capabilities approach, however, it is important to consider affordability from a social practice/social value perspective as contrasted with a simplistic income-based affordability measure (Milne, 2003).

In sum, affordable ICTs are an important consideration for this initiative because they are vital to creating conditions that expand the range of choices available to users not only in terms of access to communications services, but also in terms of how those services might be taken up and configured in ways suited to the social practices, needs and aspirations of communities, and their members.

5.2.2 Operationalizing the capabilities approach

Emphasizing capabilities as a development aim implies that individuals not only have access to affordable ICTs and services, but also that they can effectively adopt and use ICTs in ways that are responsive to their needs and ambitions. Both Kleine and Gigler have introduced theoretical frameworks that attempt to operationalize the capabilities approach. Gigler (2011, 2015), drawing on the Sustainable Livelihoods Framework, emphasizes the role of intermediaries in contributing to informational capabilities as a step toward enhanced human and social capabilities. Kleine (2010, 2013), on the other hand, adopts an empowerment perspective to introduce a Choice Framework that directs efforts toward 'achieved functionings' as a proxy for capabilities. Both offer important insights with regard to putting the capabilities perspective into practice within an ICT4D context.

We have drawn extensively on Kleine's distinction of various degrees of empowerment as they relate to choice: (i) existence of choice; (ii) sense of choice; (iii) use of choice; and (iv) achievement of choice.

These degrees of empowerment express an increasing ability to act on choice in a way that supports a variety of self-determined outcomes. With regard to ICTs, then, empowerment is a function of access, awareness, and ability with individual, community, and institutional considerations at play. For example, the *existence* of an affordable text messaging system sets the stage for an initiative where participants are made aware of a new communicative possibility and given *a sense of choice* through exposure to it. *Use of that choice* requires an individual with skills and training to configure and operate the system in an intended manner. *Achievement of that choice*, however, will require the participation of others beyond the individual, including other users in the community as well as institutional actors willing, at minimum, to permit and perhaps encourage use of the system.

The field of community informatics offers the concept of 'effective use' in a similar way to express this important consideration. Michael Gurstein first introduced the notion of effective use in 2003, describing it as '(t)he capacity and opportunity to successfully integrate ICTs into the accomplishment of self or collaboratively identified goals' (Gurstein, 2003). There is consonance between this idea and the principles of the capabilities approach.

Effective use is a multifaceted idea involving degrees of empowerment similar to those identified by Kleine. Gurstein lists seven elements, the first three of which are concerned with access to infrastructure and devices. The next two elements refer to availability of content and services suited to local circumstances. Element six mentions 'the application being presented includes provision for *capacity development* locally sufficient to its successful implementation' [emphasis in original], and element seven is the establishment of a viable, sustainable enabling framework that includes 'an appropriate structure for finance and governance' with 'active participation on the part of the local community to "animate" the process of technology acquisition and implementation'.

With a focus on affordable ICTs, our approach assumes that the existence of choice is not a major obstacle to effective use and that we can therefore direct much of our effort to processes that will lead to enhanced sense of choice, and then onward to enhanced use and achievement of choice in relation to adoption and use of ICTs.

5.2.3 Inclusive innovation and Kleine's choice framework

Assuming that community members will have access to affordable ICTs by virtue of the fact that some of these technologies are in widespread use already, suggests that we need to begin by understanding how people become aware of choice and, ultimately, how they choose to incorporate ICTs into established social practices. 'Innovation' in this sense can be described as a socio-technical undertaking that begins with the existence of choice and progresses along a continuum to an enhanced sense of choice, to effective use of that choice, and then onward to achievement of that choice as reflected in a transformation of social practices.

For example, users might be first introduced to the idea that they *could* use text messaging to improve communications with the local extension officer (i.e., an enhanced sense of choice). This alone, however, will probably not be enough to establish this choice as a communicative practice. The introduction of a text messaging

system with support from an intermediary may be necessary to encourage the use of that choice by community members. As the system comes to be used and evaluated, the community will ultimately choose whether to continue to support it or not. Achievement of that choice lies in transforming the social practice associated with livelihood communication and encouraging ongoing systemic support from other key actors.

From a capabilities theory perspective, the achievement of choice might be regarded as a culmination outcome underpinned by a more significant *comprehensive outcome* of having users directly involved in the process each step of the way. When community members are involved in this way we can also refer to it as a form of 'inclusive' innovation. Inclusive innovation is a term found within the ICT4D literature that Heeks describes as 'the means by which new goods and services are developed for and/or by those who have been excluded from the development mainstream; particularly the billions living on lowest incomes' (Heeks *et al.*, 2013). Heeks makes some further distinctions, suggesting that different aspects can be organized along a 'ladder of inclusive innovation', which progresses upward from innovations that target excluded populations toward active involvement in the processes and governance of innovation systems themselves (*see* Table 5.1). Each rung in the ladder represents a more comprehensive form of inclusiveness, building on each level as one progresses upwards.

Heeks also raises an important consideration with respect to which members of an excluded group may be included at any point on this ladder. It is likely, he suggests, that various subsets of the excluded group may be more likely to be involved depending on the level of inclusiveness. This raises questions about representation and inequality within the group that must be factored into this model. For example, the existence of cheap mobile phones and low-cost prepaid plans might be considered an inclusive innovation at Level 1 because they are marketed to the needs of a low-income market segment – a segment that might otherwise be excluded from access to telephone service. The scope of inclusiveness in this case might extend to many, perhaps all, adult members of an otherwise excluded community but of course will be also influenced by gender considerations as well

Table 5.1.   Inclusive innovation ladder (Heeks *et al.*, 2013).

Level	Inclusiveness	Description
1 (lowest rung)	Intention	An innovation is intended to address needs or wants of an excluded group. Heeks makes a point of saying that this is about motive and not necessarily 'any concrete activity'
2	Consumption	Adoption and use by the excluded group. Accessible and affordable; group members have sufficient motivation, knowledge, and skills 'to absorb' the innovation
3	Impact	An innovation has a positive impact on the 'livelihoods' of the excluded group; 'impact' can be measured in many different ways, including empowerment through choice and capabilities (i.e., Kleine and Sen)
4	Process	Members of the excluded group are involved in the development of the innovation; consideration for various sub-processes of innovation and engagement in process (e.g., being informed, being consulted, being involved in decision making)
5	Structure	Innovation processes occur within 'a structure that is itself inclusive'; institutional and organizational reforms may be required to achieve 'deep inclusion' with structure
6	Post-structure	Innovation occurs within 'a frame of knowledge and discourse that is itself inclusive'

as other social norms and practices of that group.

Further up the rungs on the ladder, at Level 4 (Process), it may be more likely to expect that a subset of the group will be at the centre of the innovation process. For example, the introduction of a crowdmapping initiative requires a certain degree of technical skills and resources that will not be available to all members of a group. In this case, there may be specific individuals more likely (or better qualified) to initiate and become involved in this process, even though the benefits of the initiative may ultimately accrue to the whole group at Levels 1, 2, and 3.

In our project we envision the role of an intermediary that spans levels inasmuch as it attempts to ensure a means for representing the interests and motives of community members in processes, structures, and possibly discourse itself (i.e., organizational policy and practice). The intermediary can play a pivotal role at Level 4 (Process) by representing and engaging with the community in the process of innovation. Similarly, the intermediary contributes to structure and post-structure levels of inclusiveness by assessing outcomes and reporting on impacts in ways that might trickle up and encourage organizational changes and challenge policy discourse to further recognize the value of inclusivity.

5.3 Technology Steward as Intermediary

In his discussion of impact evaluation of inclusive innovation Heeks suggests that, in addition to conventional economic and livelihood indicators, researchers might consider adopting:

> . . . Senian terms to research the impact of innovations on the capabilities and functionings of excluded users. [Impact assessment] could move towards more of a process approach to evaluation, seeking innovation less as a delivered good/service and more as the development of a capacity and culture within low-income communities; thus moving up the inclusive innovation scale.
>
> (Heeks *et al.*, 2013, p. 23)

Research by Gigler and others across ICT4D presents good evidence making the case for the vital role of intermediaries in creating this 'capacity and culture' for innovation within low income communities.

Our project has drawn on the Communities of Practice literature, where we have found a compelling intermediary role identified as that of a 'technology steward':

> Technology stewards are people with enough experience of the working of a community to

understand its technology needs, and enough experience with or interest in technology to take leadership in addressing those needs. Stewarding typically includes selecting and configuring technology, as well as supporting its use in the practice of the community.

(Wenger *et al.*, 2009, p. 25)

Within the context of a community of practice, a technology steward could take varied forms. For instance, it might be a part-time role taken up in response to an immediate need within the community. In another case in might be an ongoing commitment within a broader set of responsibilities. Participation of a technology steward might be voluntary and self-appointed, but within an organizational context it might still require allocating time and resources for an individual to take up the role. On the other hand, a technology steward may be assigned to the role by an organization or appointed by members of the community. The role might be active on a day to day basis, or it may only be required periodically when a particular challenge or opportunity arises – perhaps for example on a seasonal basis when a community might seek to use ICTs for a specific type of activity (e.g., sharing market access information). Motivations for taking on the role of technology steward are varied but include leadership opportunities, personal learning and growth, reputation building, and satisfaction in serving the community (Wenger *et al.*, 2009, p. 29).

One advantage to the technology stewardship approach is that it can accommodate a range of possible intermediaries suited to local conditions and preferences. Gigler (2015) identifies two basic types of roles, differentiating between 'ICT' and 'social' intermediaries. He notes that ICT intermediaries typically are from 'a specialized organization from outside the community' whereas a 'social intermediary' is situated locally, with a long-lasting relationship of trust to community members. On the one hand, ICT intermediaries are crucial for introducing and supporting new technology initiatives with a community. On the other hand, a social intermediary 'is critical for facilitating a community engagement process by which all actors within a community are invited to participate fully in the project.'

In other words, the intermediary role is necessarily multifaceted as it translates and mediates between established social practices and new possibilities. Gigler (2015) emphasizes the role as one of creating an 'enabling environment' for improving access to information through ICTs.

In this light, an appealing feature of the technology stewardship approach is its embrace of a multifaceted role in creating an enabling environment a *community of practice*. Wenger *et al.* (2002, p. 4) define a community of practice as 'groups of people who share a concern, a set of problems, or a passion about a topic, and who deepen their knowledge and expertise in this area by interacting on an ongoing basis'. This provides an important consideration because in some cases motivation for adoption and use of ICTs might be situated more clearly with respect to specific professional or livelihood concerns than local geographical considerations. For example, farmers that share concerns about a crop variety or particular cultivation technique may have specific informational needs that cut across geographical regions. A technology steward situated within this group would direct efforts not at a specific locale but rather at creating an ICT enabling environment for a community of practitioners that identify with this domain of interest.

The steward's role is essential to help create an enabling environment that will support ICT adoption and use in relation to the various activities and aspirations of a community of practice. The approach laid out in Wenger *et al.* (2009) provides a useful blueprint for designing a technology stewardship training program that incorporates community assessment and engagement, as well as technology acquisition and implementation. Through the lens of Kleine's Choice Framework, the technology steward can be seen to serve four primary responsibilities with regard to the adoption and use of ICTs:

- Make the community aware of the existence of choice.
- Help the community to develop a clear sense of choice.
- Facilitate and support the effective use of choice.
- Recognize and sustain the achievement of choice.

These responsibilities are carried out through an ongoing process of community engagement that involves a set of activities including community engagement, rapid prototyping with low-cost ICTs, limited duration campaigns, and participatory evaluation and assessment.

5.4 Piloting Technology Stewardship in Sri Lanka

Between 2012 and 2015, the University of Alberta, Wayamba University of Sri Lanka, LIRNEasia, and the University of Guelph were involved in a partnership development project funded by the Social Sciences and Humanities Research Council of Canada (SSHRC). A primary objective of the project was to establish a community–university research partnership to explore the potential for low-cost ICTs to enhance knowledge mobilization practices within agricultural communities of practice in Sri Lanka (Gow et al., 2015).

Through a series of workshops and consultations held in Sri Lanka between 2012 and 2014, an agreement was eventually established between the university research team, the Sri Lanka Department of Export Agriculture (DOEA), Rangiri Radio, and the non-governmental development organization Janathakshan to conduct a set of four small-scale pilots, or 'communication campaigns' at four different locations in Sri Lanka using an action research approach with active collaboration between researchers and participants (Jayathilake et al., 2015) (see Table 5.2).

As part of the community consultation process, a technology steward was nominated by each of the sponsoring organizations. The DOEA nominated Extension Officers who were interested in using text messaging to improve the timeliness and lower costs of notifications to and from the community; a community liaison officer from Janathakshan was interested in using text messages to improve market price sharing for the community; and a staff member from Rangiri Radio was nominated to work within the organization and integrate FrontlineSMS[2] into existing broadcast programming in order to encourage listeners to interact with text messages during farm radio segments.

All of the appointed technology stewards were trained by a member of the research team from Wayamba University to use and customize the FrontlineSMS software platform and were also introduced to some basic participatory research methods intended to help them engage with their communities in both promoting and sustaining the use of the new services that they would be introducing with their communities during the campaign. Data were collected at various times during the campaigns using a mixed-method approach including semi-structured interviews with four technology stewards and their sponsoring organizations.

The technology steward in the DOEA North campaign was an Extension Officer working the area and familiar to the community. He proved to be a keen user of the system and sent over 700 messages over the course of several weeks, providing farmers with reminders of upcoming meetings, information on crop disease and other best practices. However, relatively few farmers used the system to reply or ask questions. Further efforts by the technology steward to encourage input from farmers were constrained by their reluctance to compose text messages (as contrasted with reading messages sent to them).

Table 5.2. Summary of communication campaigns conducted in Sri Lanka for the pilot study.

Sponsor	Farming practices	Campaign focus	Technology
Janathaskhan (Eastern Prov.)	Subsistence farming/ fishing	Market price sharing	Mobile phones/Frontline SMS
DOEA-North (Central Prov.)	Ginger farmers	General notifications from Extension Officer	Mobile phones/Frontline SMS
DOE-South (Central Prov.)	Pepper farmers	General notifications from Extension Officer	Mobile phones/Frontline SMS
Rangiri Radio (Nationwide)	Broadcaster	Agricultural programming	Mobile phones/Frontline SMS

In consultation with the community, the technology steward suggested to the research team that a voice-based system might overcome some of these barriers, and the project team is now working with DOEA to launch a follow-up campaign that will include Freedom-Fone (a voice-based messaging system).

The DOEA South campaign underperformed when compared to its counterpart in the North. In the DOEA South campaign the technology steward role was shared between two Extension Officers, resulting in a split in responsibilities. Our initial observations suggest that this made it difficult for either steward to coordinate their actions and to develop a sense of ownership of the campaign, as one steward was responsible for managing FrontlineSMS, the other for prototyping with the community.

The Rangiri Radio campaign received full support of the station management who saw it as a way to enhance its radio programming. Listeners were asked to text in song requests or other questions. The software records show hundreds of incoming messages each day of the campaign, which pushed the capabilities of FrontlineSMS to its limits (Rashid *et al.*, 2016). Curiously, the technology steward did not initially use the system for their farm radio program, instead opted to introduce it with their popular music shows, which of course opened up the question concerning situations in which a technology steward uses a system in ways not initially intended by the sponsor or the research team. The team discussed this situation and decided that if the use was generally consonant with the wider community of interest, it is important to let the technology steward explore innovation in ways that may not immediately respond to a particular need but instead lay the ground to prepare the community for the introduction of other campaigns. In this case, the Rangiri Radio campaign was seen to be offering its listeners a new form of interactivity in a compelling format (song requests) that could then be subsequently introduced into its farm radio programming.

In the case of the Janathakshan campaign, the technology steward had comparably little support from the sponsoring organization, in part due to changes that took place with the sponsor during the campaign. However, he was able to send out a large number of messages initially in an effort to get farmers to self-subscribe to the system but had modest uptake from the community. This outcome may be partly explained by survey data that we collected on use and adoption of technology in the community that showed a relatively low level of prior use of text messaging. This was compounded by social and economic challenges stemming from post-war conditions in this northern region of Sri Lanka, as well as literacy barriers for Tamil speakers trying to understand messages composed in phonetic Tamil using the limited character set available on their mobile phones.

With the conclusion of the campaigns in August 2014, both DOEA North and Rangiri Radio expressed interest in continuing with the project and expanding beyond FrontlineSMS to begin experimenting other ICT platforms, including an open-source interactive voice response system. Technology stewards in both cases remain actively involved as they continue to liaise with their communities. In the less successful cases, our initial analysis suggests that we need to examine more closely two key considerations: if and how technology stewardship can be effective as a shared responsibility within a community, as in the DOEA South campaign; and to what extent a technology steward may need to engage with and prepare a community for the introduction of a new service (e.g., basic technology literacy workshops), as in the Janathakshan campaign. This kind of training can also be helpful to identify and address unforeseen systemic issues in the adoption and use of the technology by the community.

Project partners also concluded that there could be longer-term benefits if we were to introduce a more formal training program for technology stewardship that expanded on the skill set that was provided in the ad hoc training sessions for the pilot research. Moreover, the DOEA expressed interest in such a training program as a step toward scaling up the campaigns to include Extension Officers in other districts. As such, we believe there an opportunity to develop, deliver, and evaluate a technology stewardship training program that addresses both technical and community engagement aspects in its curriculum, with a view to building a cohort of technology stewards in the agricultural sector who can enhance the range of ICT choices available to their communities of practice.

5.5 Putting Technology Stewardship into Practice

Discussions that followed from the pilot studies resulted in the project being awarded additional funding in late 2015 to develop and test a technology stewardship training workshop as part of a Joint Education and Training Initiative (JETI).

The JETI is proposed to serve not only to support technology stewardship training but also to provide experiential learning opportunities for post-secondary students in Canada and Sri Lanka as part of an action research collaboration. In other words, the JETI is an overarching initiative intended to serve multiple roles: (i) it is a training program in technology stewardship for practitioners that will provide direct benefit to those practitioners and their respective communities of practice; (ii) it is a vehicle by which we can continue to conduct action research on technology stewardship and inclusive innovation with these communities; and (iii) it creates a unique learning opportunity for post-secondary students interested in ICT4D and Community Informatics.

Based on findings from the initial pilot work, the team has completed an initial design of curricular materials with a set of four primary learning outcomes:

- Identify and describe the key principles and activities involved in technology stewardship as a leadership role.
- Lead a community engagement activity to identify a key ICT need or priority for a limited duration communication campaign with your community of practice.
- Plan, design and implement a technology prototype and related activities needed to carry out a limited duration campaign using a low-cost ICT platform.
- Plan, design and implement an evaluation plan to assess and report on the outcome of the communication campaign and your involvement as a technology steward.

The practitioner curriculum is designed around a set of four learning modules, intended to be delivered in a set of three-hour sessions over a two-day period. The post-secondary curriculum will be enriched with readings from theory and empirical studies in each of the four areas.[3]

5.5.1 Measuring outcomes and understanding impact

Looking beyond the immediate goals of the training program we have proposed, there is of course the looming research question that remains as to the effectiveness of the technology stewardship model in achieving its objectives as an ICT4D initiative.

Here we have developed a preliminary analytical framework that combines Heeks' ladder of inclusive innovation as degrees of empowerment through 'achieved functionings' as suggested by Kleine's Choice Framework. Gigler's 'ICT Impact Chain' (Gigler, 2011) provides further guidance with regard to detecting the presence of 'enhanced informational capabilities' as expressed through specific indicators.

The impact of a technology steward within a particular community of practice can be assessed by examining the impact of a campaign as a deliberate intervention with a specific intended outcome. Here we are interested in what ways and to what degree the campaign has resulted in meaningful use of the ICT in relation to the intended outcome. In other words, during the campaign did the community demonstrate enhanced informational capabilities measured by indicators, such as such as improved capacity to use different forms of ICTs, enhanced information literacy, enhanced capacity to produce and publish local content, improved ability to communicate with others in relation to the campaign objectives? This corresponds with levels 1–3 on the Inclusive Innovation Ladder and reflects a fairly narrow assessment of 'impact' that does not necessarily include wider social or economic development objectives.

Even if a campaign impact is limited, it may nevertheless lead to significant changes in capabilities among community members in terms of their perception of choice. As such we need to inquire as to how and to what degree the technology steward's presence might have resulted in an enhanced sense of choice within the community with regard to ICT use. This is an important distinction because it is anticipated that some campaigns will not achieve their immediate objective but may still produce positive outcomes by involving community members in the planning process, and by enhancing a sense of choice within the community, while perhaps even generating other ideas for applications of ICT. This corresponds with level 4 and touches on level 5 on the inclusive innovation ladder by assessing the contribution of the technology steward and the campaign structure as a means to engage community members in a process of innovation.

The anticipated timeline for assessing campaign level impacts such as these falls within the range of several weeks to a few months, depending on the length of the campaign. Results can be fed back to the technology steward and community in an iterative manner typical of action research.

At a program level, the goal of a proposed Joint Education and Training Initiative is to better understand and improve technology stewardship through a program of research and practitioner training. The impact of the initiative is measured as a collective outcome of the individual technology stewards working with their communities but also by the possibility of bringing into existence a community of practice of technology stewards unto itself as a collection of individuals involved in that role and engaging with their communities. Such a community could be expected to help support and sustain an active group of technology stewards embracing the principles and processes of inclusive innovation and carrying these forward to user communities through campaigns and other activities.

Program-level outcomes such as these might be expected to take several years to assess but ultimately would aim to provide insight as to the contribution of the technology stewardship model to enhanced informational capabilities as they relate to a wider set of 'achieved functionings' and development goals.

5.6 Conclusion

This chapter has outlined a technology stewardship model being developed and tested by a community–university research partnership with Canadian and Sri Lankan organizations. The model is grounded theoretically in the capabilities approach as operationalized through an integrated framework informed by several perspectives including Kleine's Choice Framework, Heeks' ladder of inclusive innovation, and Gigler's concept and impact assessment model based on 'informational capabilities'.

Results from a pilot study using a technology stewardship model have led us to begin to develop a formal training program that includes techniques in community engagement, rapid prototyping, and evaluation and impact assessment. With the completion of an initial training workshop in Sri Lanka in September 2016, we intend to expand the program going forward to include a post-secondary course that will create opportunities for collaborative cross-cultural learning opportunities in ICT4D and inclusive innovation.

Results from the work are expected to appear in several stages, beginning with short-term assessment of campaign outcomes and leading to more comprehensive program measures as the program grows and matures.

Acknowledgement

This research was supported by the Social Sciences and Humanities Research Council of Canada.

Notes

[1] Our initiative focuses on the adoption and use of digital technology but we are also mindful of how individuals and communities might continue to value analog technology as part of their social practices and, moreover, we consider the potential of digital and analog systems working together as a form of hybrid media (e.g., the use of traditional (analog) broadcast radio in combination with text messaging on mobile phones).
[2] See: http://www.frontlinesms.com/ (accessed 1 March 2018).
[3] Working with the DOEA and other partners we have recruited 18 participants from that organization to attend the course to be offered in late September. The course will provide an opportunity to test the curriculum, receive feedback from participants, and to continue to revise and refine our approach to technology stewardship training within the context of an ongoing action research project.

References

Duncombe, R. (2012) Mobile phones for agricultural and rural development: a literature review and future research directions. *Development Informatics Working Paper No 50*, University of Manchester, UK.

Farm Radio International (2011) How ICTs are changing rural radio in Africa: the new age of radio. Available at: http://www.farmradio.org/pubs/farmradio-ictreport2011.pdf (accessed 14 November 2013).

Foster, C. and Heeks, R. (2013) Conceptualizing inclusive innovation: modifying systems of innovation frameworks to understand diffusion of new technology to low-income consumers. *European Journal of Development Research* 25, 333–355.

Foster, C. and Heeks, R. (2014) Nurturing user–producer interaction: inclusive innovation flows in a low-income mobile phone market. *Innovation and Development* 4, 221–237.

Gigler, B.S. (2011) Informational capabilities: the missing link for the impact of ICT on development. Available at: http://dx.doi.org/10.2139/ssrn.2191594 (accessed September 2017).

Gigler, B.S. (2015) *Development as Freedom in a Digital Age.* World Bank, Washington, DC, USA.

Gow, G., Waidyanatha, N., Jayathilake, C., Odame, H.H., Barlott, T. *et al.* (2015) Fostering inclusive innovation for agriculture knowledge mobilization in Sri Lanka: a community–university partnership development project. 7th International Conference on Communities and Technologies, University of Limerick, Ireland, 27–30 June 2015

Gurstein, M. (2003) Effective use: a community informatics strategy beyond the digital divide. *First Monday*, 8. Available at: http://firstmonday.org/article/view/1107/1027 (accessed 9 May 2017).

Heeks, R., Amalia, M., Kintu, R. and Shah, N. (2013) Inclusive innovation: definition, conceptualization and future research priorities. *Development Informatics Working Paper No 53*, University of Manchester, UK.

Jayathilake, H., Jayasinghe-Mudalige, U., Gow, G., Waidyanatha, N. and Perera, I. (2015) Use of low-cost information and communication technologies for knowledge mobilization in agricultural communities in Sri Lanka. 8th International Research Conference, General Sir John Kothalawala Defence University, 27–28 August 2015, Rathmalana, Sri Lanka.

Kleine, D. (2010) ICT4what? – Using the choice framework to operationalize the capability approach to development. *Journal of International Development* 22, 674–692.

Kleine, D. (2013) *Technologies of Choice? ICTs, Development, and the Capabilities Approach.* MIT Press, Cambridge, Massachusetts, USA.

Milne, C. (2003) Measuring affordability of telecoms, 3rd World Telecommunication/ICT Indicators Meeting. International Telecommunications Union, Geneva, Switzerland.

O'Donnell, A. (2011) 'Farming out' agricultural advice through radio and SMS. Available at: http://voices.nationalgeographic.org/2011/04/26/%E2%80%9Cfarming-out%E2%80%9D-agricultural-advice-through-radio-and-sms/ (accessed 14 November 2013).

Qiang, C.Z., Kuek, S.C., Dymond, A. and Esselaar, S. (2011) *Mobile Applications for Agriculture and Rural Development.* ICT Sector Unit, World Bank, Washington, DC.

Rashid, F., Hambly-Odame, H., Gow, G., Waidyanatha, N., Jayasinghe, U. and Hack, J. (2016) Converging frontlineSMS, Freedom-Fone and radio for mobilizing knowledge for sustainable agriculture in Sri Lanka. 4th Annual International Conference on Sustainable Development (ICSD), 21–22 September 2016, Columbia University, New York, USA.

Waidyanatha, N., Gow, G., Barlott, T. and Jayathilake, C. (2015) Technology stewardship, text messaging, and collaboration in agricultural work: preliminary results from an action research study in Sri Lanka. Computer Supported Collaborative Work Conference, 14–18 March 2015, Vancouver, Canada.

Wenger, E., McDermott, R. and Snyder, W. (2002) *Cultivating Communities of Practice.* Harvard Business School Press, Boston, Massachusetts, USA.

Wenger, E., White, N. and Smith, J.D. (2009) *Digital Habitats: Stewarding Technology for Communities.* CPsquare, Portland, Oregon, USA.

6 Reducing Transaction Costs in Contract Farming Arrangements: the Case of Farmforce

Fritz Brugger*

NADEL Center for Development and Cooperation, ETH Zurich, Switzerland

6.1 Introduction

Smallholder farmers produce less on a hectare of land compared to professional farmers under comparable conditions. Too many farmers don't even produce enough to work themselves out of poverty. A combination of lack of technology, know-how and access to markets, together with ill-informed agricultural policies is typically seen as holding back the 525 million smallholders worldwide (IFC 2013). Although some seem to perceive smallholders as an anachronism there is no massive move out of rural areas and out of agriculture imminent (Collier and Dercon 2014). Rather, smallholder farmers can be expected to play an important role in producing the agricultural crops required to meet the duplication in demand as the world's population reaches 9.1 billion by 2050 (IFC 2013).

Improving smallholder farmer agriculture has experienced renewed attention by development agencies and policy makers. Reaching the huge number of smallholder farmers working in remote rural areas often is the single biggest problem. This 'last mile' to the farmers' gate – or rather the 'first mile' of the food value chain – is where many well-intended projects fail. Mobile technology is increasingly seen as a major opportunity to bridge this gap (World Bank 2016). Optimistic calculations expect additional income of US$220bn until 2020 for smallholder farmers from mobile technology services, mainly improved access to financial services, access to agricultural information, improved data visibility for supply chain efficiency, and enhanced access to markets (Vodafone, 2011).

6.2 Contract Farming

In this chapter, we focus on contract farming as one approach to support smallholder agriculture to discuss the above-mentioned 'first mile' challenge and how to address it effectively. Contract farming can be an effective institution for helping small farmers raise their productivity and orient their production toward more remunerative commodities and markets. Yet, contract farming cannot serve as a broad-based strategy for rural development; it only makes economic sense for certain commodities in certain markets. Contract farming is most frequently used in the production of high-value crops for domestic formal markets and for export (Minot, 2007).

Contract farming can be defined as an agreement between farmers and processing and/or marketing firms for the production and supply of agricultural products under forward agreements. The agreement frequently predetermines the prices. Further, the arrangement

---

* E-mail: fritz.brugger@nadel.ethz.ch

involves the purchaser in providing a degree of production support through, for example, the supply of inputs (which *de facto* is a production credit) and the provision of technical advice. In sum, the basis of contract farming arrangements is: (i) a commitment on the part of the farmer to provide a specific commodity in quantities and at quality standards determined by the purchaser; and (ii) a commitment on the part of the company to support the farmer's production and to purchase the commodity (Eaton and Shepherd, 2001).

A broad range of contract farming models exist (for a discussion of the basic arrangements, see GIZ 2008). According to Eaton and Shepherd (2001), the intensity of the contractual arrangement varies according to the depth and complexity of the provisions in each of the following three areas:

- Market provision: the grower and buyer agree to terms and conditions for the future sale and purchase of a crop or livestock product.
- Resource provision: in conjunction with the marketing arrangements the buyer agrees to supply selected inputs, including on occasion land preparation and technical advice.
- Management specifications: the grower agrees to follow recommended production methods, inputs regimes, and cultivation and harvesting specifications.

Empirical studies consistently support the positive contribution of contract farming to production and supply chain efficiency. A recent systematic literature review of 11 studies on the impact of contract farming on productivity, and 12 studies on the effects of contract farming on farming income revealed that almost all the selected studies of the latter category argue that farmers on contract farming schemes experienced some increase in their income (Bellemare, 2012; Wang *et al.*, 2014; Nguen *et al.*, 2015).

However, participation in contract farming *per se* is not a guarantee for increasing farmer income. It is well recognized that contractual design factors are important determinants of the welfare impacts of the participation in the contract (Abebe *et al.*, 2013; Fullbrook, 2014).

Several factors of the contract design and execution seem to be particularly relevant. First,

the contract design introduces a selection effect for participation: in contracts where inputs or interlinked services are not provided by the contractor as in-kind production credit, credit lines from banks or microfinance institutions (MFI) play a significant role in alleviating liquidity constraints that characterize smallholder systems. Households that have access to credit are more likely to invest in order to meet the buyer's quality requirements, which can earn them premium prices. Second, the distribution of the production and marketing risks between the farmer and the buyer has a significant bearing on whether the contract results in higher farmer income; this includes terms of the contract but also access to information to all parties involved (Mwambi *et al.*, 2016). Third, experience at the operational level shows that the quality of the business information system in place (be it analogue or digital) has a major bearing on the success of a contract farming scheme (GIZ 2008).

Adherence to voluntary sustainability standards (VSS) is a second mechanism through which smallholder farmers can get access to formal markets and – at least in some instances – benefit from a premium for their produce. VSS are private governance initiatives to shape global supply chains with the aim to facilitate more socially and environmentally responsible behaviour (Gereffi *et al.*, 2001). VSS vary in scope, i.e. the extent to which they emphasize labour conditions; economic productivity and environmental issues in requirements, i.e. how demanding and strict the rules are defined; and in enforcement, i.e. the design of the conformity assessment from self-declaration to third-party attestation. Voluntary sustainability standards have grown rapidly in number and importance in global commodity markets over the past decade. The average annual growth rate of standard-compliant production across all commodity sectors in 2012 was 41%, significantly outpacing the annual average growth of 2% in the corresponding conventional commodity markets. Sustainability standards have penetrated several mainstream commodity markets. For example, standard-compliant coffee reached a 40% market share of global production in 2012 (up from 15% in 2008), cocoa moved from 3 to 22%, palm oil from 2 to 15%, and tea from 6 to 12% in the same period (Potts *et al.*, 2014).

The industry sponsored Global Partnership for Good Agricultural Practice (GLOBALG.A.P.) standard is an example of a voluntary certification scheme that is increasingly defining market access for smallholders in developing countries. Founded in 1997, the GLOBALG.A.P. is a private initiative operating in the food and agriculture sector across 110 countries. GLOBALG.A.P. has become the *de facto* standard for horticulture exports to Europe and the USA. A study into pineapple farming in two pineapple growing districts in Ghana found that 70% of GLOBALG.A.P. certified pineapple farmers have access to formal markets (processors and exporters) compared to 30.3% of non-certified farmers. Regarding the economic performance, GLOBALG.A.P. certified pineapple farmers obtained 2.4 times more net average income than non-certified pineapple growers. This is primarily the result of a 33.5% higher productivity of certified farmers while the average sales prices across all sales channels was only 5% higher for certified pineapple farmers (Kuwornu and Mustapha, 2013).

6.3 Voluntary Certification Schemes

Often, contract farming and compliance with certification schemes are mutually dependent: for farmers, VSS are a means to access markets. Kariuki (2014) found that successful transition to GLOBALG.A.P. certification depends on training, the farm asset base, and organized production. On the other side, the lack of money to pay for certification and audit are constraints hindering complying with GLOBALG.A.P. standards. For a third of the certified farmers, raising money to pay for external audits and the proper keeping and storage of past records are difficult; two thirds struggle with keeping the rules and procedures consistent. For farmers, food safety requirements, the quality demanded, and social and environmental standards are often as much an entry barrier to formal markets as they are an opportunity to earn a stable income (Mumo, 2012; Okello, 2015).

For processors, retailers and exporters (collectively called 'aggregators' in this article) sourcing from smallholder farmers, VSS are a means to standardize quality and to communicate the level of quality provided downstream the value chain to business partners and consumers.

Yet, deploying the harmonizing and tools of certification schemes only adds to the fact that sourcing from smallholder farmers is an operational headache. Be it horticulture, fruits and vegetables, flowers, rice, soybeans, potatoes, cocoa, cotton or any other produce that is suitable for contract farming arrangements: the requirements for consistent quality, predictable quantity and traceability translate into high transaction costs, i.e. the cost for the facilitation (identifying and assessing farmers), conclusion (contract negotiation and design) and execution (distribution of inputs, technical assistance to farmers, management and logistics) of the contract as the cost incurred for the control of the quality produced and adaptation of the contract farming arrangement. As a result, aggregators often work with far fewer farmers than might be possible and prefer to rely on highly mechanized captive farms where few local farmers find seasonal jobs as labourers, e.g. for harvesting (Bijman, 2008). The high transaction cost involved in contract farming puts limits to a model that also has a positive effect on farmer welfare and productivity (Bellemare, 2012; Wang *et al.*, 2014; Nguen *et al.*, 2015).

6.4 Mobile Technology to Manage Contract Farming Arrangements

Against this background, Farmforce was created as a mobile-phone-based Software-as-a-Service (SaaS) business management system that makes the farm–firm interaction more effective and efficient. The theory of change informing the development of Farmforce is that the reduction of transaction cost in the form of: (a) improved efficiency of farmer identification, coordination and interaction; (b) comprehensive and real-time availability of management information; (c) improved agronomy; (d) reduced cost of monitoring standard compliance and standard audit; and (e) simplified downstream value chain coordination will make contract farming more effective and efficient, and hence a more attractive production option for aggregators compared to producing in captive farms. In turn, this will lead aggregators to involve more farmers in contract farming arrangements though which they get access to formal markets and earn more income. The theory of change holds under the condition that the envisaged technology solution is: (a) scalable; (b) applicable across different value chains; and (c) affordable for the end user of the software solution.

The role of technology in the theory of change is very clear: to enable the reduction of transaction cost as specified in (a)–(e) on p. 56 and again (a)–(c) above.

However, before calling a software company to get on the task, it is worth considering prior experience with mobile technology projects that attempt to improve smallholder farmer productivity in one way or another. Many development agencies have ventured into this area with mixed success. An in-depth stocktaking of mobile applications for agriculture as well as the experiences gained with the technology has produced a number of relevant insights that help to avoid the most common pitfalls of ICT4D projects (Brugger, 2011):

- First, *avoid a narrow focus*. Mobile technology projects often emanate from particular projects with the aim to solve one specific issue. Scaling to other topics or to more users is put aside until the initial challenge is solved and the pilot works. Also, mobile technology projects are often project managers' or organizations' pet projects. As a consequence, technology and design choices are made that most likely will impede efficient scaling up.
- Second, *develop a business case*. In almost all cases reviewed, the financial business was driven by donors or by research and reliant on donor funding. No one had a clear concept as of how to move from pilot to scale. The lack of a working revenue model puts the sustainability of any mobile technology offer into question. This also holds for solutions that build on open-source software; even in this case a business model is needed to cover ongoing operational and maintenance cost (Schireson and Thakkar, 2016).
- Third, *be aware of the technology adoption curve*. It points to the chasm between early adopters of technology and the majority of the intended end users (Rogers, 1983;

Moore, 2014). Most often, reports about pilot deployments of mobile technology for smallholder farmers are enthusiastic and conclude with a positive outlook for scaling up. Yet, it turns out that farmers don't use the mobile technology solutions at scale beyond the pilot group. The reason behind this is that early adopters tend to be over-represented in pilot groups which are not representative of the large majority of the envisaged end-users. Early adopters are more open to technology and more willing to try out new ways of doing things. More recent research has confirmed the difference between pilot groups buying into mobile technology solutions and the majority of farmers staying away from technology.

The lessons illustrate the need to think about scaling-up as well as about a sustainable business model early on, since this will directly influence the concept and architecture of the software. In order to avoid the pitfalls identified, the Farmforce strategy is based on the following business model.

6.4.1 The Farmforce business model

Although the motivation and goal behind Farmforce is to provide more farmers access to formal markets through outgrower arrangements, the software is not designed to be used by smallholder farmers. Rather, Farmforce is built for those actors in the value chain who have a direct economic interest in reducing transaction costs involved in contract farming arrangements. Typically, this is the aggregator, be it a processor, exporter, cooperative, a nucleus farm or similar. This economic interest of the aggregator translates into the willingness to pay for a technology solution (provided it meets their expectations), and – equally important – it is the best guarantee that the software is properly introduced and used by the field officers (mobile application) and office staff (web application). Moreover, aggregators often have enterprise resource planning (ERP) systems or other downstream software in use; integrating with them allows to build gapless traceability from the smallholder's farm to the consumer's fork.

In such a business model, smallholder farmers will indirectly benefit from reduced transaction costs in two ways: first, they can get higher prices for their products and, second, more farmers can be involved in outgrower schemes due to gains in efficiency.

Serving hundreds of contract farming clients in developing countries with the same software is only possible when the solution is able to work: (a) across a large number of different value chains; (b) different certification schemes; and (c) different organizational set-ups. As a consequence, Farmforce cannot be a bespoke application or a solution that is tailor-made for one particular contract farming arrangement, one particular client (irrespective of how big the client is), or one value chain. Rather, it must be a product that serves hundreds or thousands of customers, similar to Salesforce that serves customer relationship management (CRM) requirements for many different businesses.

The strategic decision to take a product approach instead of a consulting approach has far-reaching consequences for the architecture of the solution. It determines how to select and design features to be implemented and how to respond to customer requests (Cagan, 2008). The core challenge of a product approach – and main difference to a bespoke solution – is that it requires finding generic ways to implement features to any given problem. The example of documenting compliance with certification standards (VSS) well illustrates this point: the permanently changing and growing landscape of VSS and the fact that most of the VSS keep changing their criteria make it impossible to build all voluntary sustainability standards into Farmforce with reasonable effort; it would be an endless maintenance hassle (*see*, for example, www.standardsmap.org for an overview). Moreover, many contract farming systems have multiple certifications with sometimes significant overlap in the criteria. The generic solution to the problem is the identification of the basic structure of VSS. It turns that any VSS is always a combination of: (a) documentation of some sort of production-related activity along the growing cycle; and (b) assessment of social or environmental or organizational conditions at the level of a farm, a field or a farmer organization that are specific to a given VSS. Analytically, an assessment is a survey whereby each survey question is ranked against some sort of criteria and the sum of the ranked questions translates into a results statement that follows a more or less differentiated pass/fail logic. Hence, building the capability to document growing activities and carrying out assessments linked to fields, farms, farmers or farmer groups allows capture of the information required for any VSS or combination of VSS and adaptation to any changes that inevitably will occur. The assessment logic can further be used for the assessment of a farmer's eligibility to join a scheme or to establish whether he/she qualifies for a loan.

The product approach allows management of the full range of outgrower arrangements and offering Farmforce under the Software-as-a-Service (SaaS) concept are key preconditions for a successful scaling-up. Therefore, the set of functionalities built into Farmforce were identified based on a thorough analysis of the organizational contract farming particularities and in close cooperation with projects that display the most complex organizational arrangements.

In order to meet the requirements of aggregators, who are Farmforce's primary user group, development of the software was conducted in close collaboration with Kenya Horticulture Exporters Ltd, a horticulture exporter in Kenya with longstanding experience with contracting smallholder farmers located in the Laikipia area. We have deliberately selected french beans production for export to the European Union (EU) as a primary test case. This is because the combination of a perishable crop produced in short cycles of six to ten weeks combined with the EU's highly regulated and strict food safety certification requirements, and strict minimum residual limits (MRL) for chemicals, represents the most complex and challenging case in managing smallholder contract farming. Over a period of two years, the partner company's chief agronomist participated in the requirements design while a team of field officers tested the application under real-world conditions. It was important to understand the various complex processes and requirements the envisioned management software should be able to handle. Accordingly, the requirements were translated into generic solution concepts and checked against a diverse set of other contract farming arrangements.

Technically, Farmforce was created following the agile development methodology: new features were regularly deployed in a number of test sites in order to make sure that the software architecture was flexible enough to allow the deployment of Farmforce across the whole range of agriculture contract farming arrangements. Further, it generated a constant stream of user feedback that greatly helped to optimize usability and process flows.

6.4.2 Farmforce concept and features

Conceptually, the core of the Farmforce architecture builds the *'virtual farm'*. This means that Farmforce models each participating farm in some detail: while incorporating the farmer's personal details and his or her affiliation with farmer groups are a no-brainer, Farmforce models each field a farmer tills (including information on ownership status) with a GPS point or – alternatively – with the geo-coordinates of each corner of the field based on which Farmforce automatically calculates the field seize. This static information about a farm's core productive assets builds the basis to model the farming activities. On any given field, Farmforce models each individual growing cycle including all activities performed from land preparation to sowing, weeding, fertilizing, transplanting, applying fertilizer or whatever activity needs to be recorded. The definition of an 'activity' is fully configurable including information on who performs the activity, recommended dosage, waiting periods, cost per unit of input (e.g. seeds, fertilizers, chemicals) and the amount of input effectively used. To make Farmforce easy to use in the field, all meta-information on inputs and activities is defined on the Farmforce web-application by the office staff so that field staff can select entries from drop-down menus and all relevant processing happens automatically.

Along the growing cycle, field officers can record yield forecasts in Farmforce at relevant growth stages (e.g. germination, flowering or podding); the number and definition of growth stages are fully configurable. Following standard agricultural practice, yield forecast is expressed as a percentage of the full harvest potential under optimal conditions (e.g. a 90% germination rate or a 75% flowering rate). Since the

information on optimal productivity is deposited as meta-data of the seed information, Farmforce calculates the harvest amount to be expected using the field size information and the information about the crop's performance at time t_1 and updates the information after the next yield forecast at time t_2, etc. As proof of the field officer's assessment he/she can add a picture of the field to the forecast and complement this information with the expected harvest date in case this deviates from the standard harvest date (calculated as sowing date plus days to maturity taken from the seed meta-information).

At the end of each growing cycle, harvesting information can be added for one full harvest or for several partial harvest rounds together with information on quantity, quality harvested, and the price paid to the farmer. To improve accuracy and transparency, an electronic scale can be connected to the mobile Farmforce device via Bluetooth and the reading of the scale is automatically entered into Farmforce. A mobile printer, also connected via Bluetooth, can print receipts at the farm-gate to be handed out to the farmer. To enable traceability Farmforce creates a unique code for each harvest batch which can also be printed out on the spot. Alternatively, Farmforce scans pre-produced QR codes or barcodes (which are more robust that just a piece of printed paper) using the mobile phone's camera. Each harvest batch can be tracked as it moves down the value chain, even as one or more larger consignment(s) and also if later split into smaller units again.

While the 'virtual farm' provides all relevant information about an individual farmer participating in a contract farming scheme, aggregators are as much interested in understanding the overall performance of a particular crop that farmers are growing collectively. For example, assume an exporter who has to honour a contract to supply 120 tons of French beans. To produce this amount, the exporter has contracted 500 farmers and he needs aggregated information across the 500 farmers' growing activities; browsing through the records of 500 farmers is cumbersome and still does not provide the information required. For this reason, Farmforce has introduced a second pillar in its architecture, which is the *'planting campaign'*. This feature allows linkages between an unlimited number of fields into one planting campaign.

(i.e. the production of 120 tons of French beans to be supplied to client x at time y taking our example above). Farmforce handles all fields linked to a planting campaign as one big field by aggregating the information of the underlying individual fields such as yield forecast or harvest information. A 'planting campaign' lasts for one growing cycle only and any field can be part of a different planting campaign during the next growing cycle.

These two core concepts – *virtual farm* and *planting campaign* – of the Farmforce architecture are the backbone to track the flow of goods, information and money (*see* Fig. 6.1).

Flow of goods

Monitoring the flow of the agricultural produce using traceability codes allows an uninterrupted chain of custody from the field to the fork. However, there are often additional goods that need to be tracked, as well such as inputs or seeds in order to guarantee and document its proper use and disposal (e.g. for chemical inputs or to make sure that that valuable items do not disappear are being misused). To this end, Farmforce includes an inventory app that allows the management of decentralized warehouses fully integrated with the functions of the virtual farm discussed above. For example, inputs used for growing activities can be automatically deducted from the warehouse stock; information on disposal of the remaining balance can be entered as well providing a full audit trail. Such a real-time overview over stock levels in decentralized warehouses and the tracking of agricultural produce and other goods simplifies logistics planning and minimizes capital outlay.

Flow of information

In addition to the information captured along the growing cycle and downstream of the value chain, the survey and assessment module discussed above allows capture of any additional information that might be required. On top of that, a separate module facilitates the management of farmer training activities including support for the monitoring and evaluation of training effectiveness. Information is not only relevant for managing contract farming, but also for other partners involved, mainly but not limited to auditors of VSS schemes and downstream processors or buyers of agricultural produce. Since those players in the value chain typically have their own IT systems in place, Farmforce offers standardized application programming interfaces (APIs) in order to seamlessly integrate with existing systems.

Flow of money

Farmforce accounts for the fact that contract farming always includes some sort of pre-financing by providing an accounts module to track all items provided as pre-financing in cash or in-kind, as well as all repayments either in cash or in-kind at harvest or at any other point in time. This financial information is also accessible by the individual farmer by sending a text

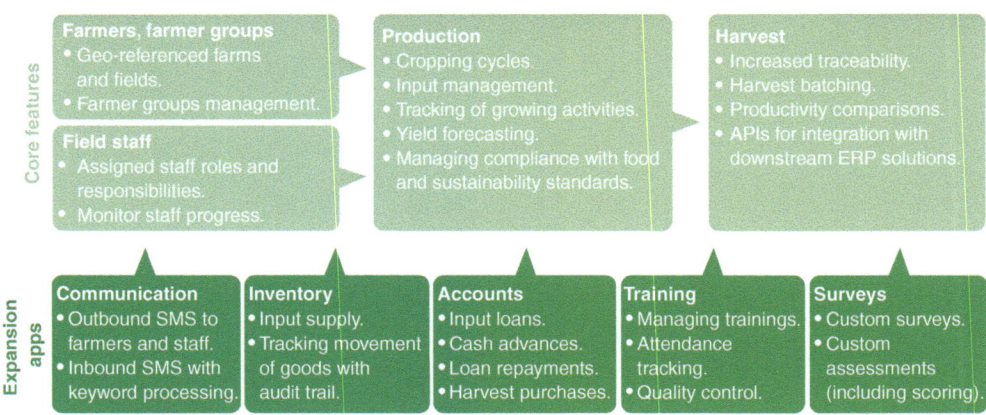

Fig. 6.1. Farmforce features.

message to his Farmforce account to query the balance and latest transactions.

6.4.3 Digital management

Farmforce generates all necessary data for steering the complex interfaces between farmers and farmer groups, field officers and famers, farm supply and off-take and to facilitate the corresponding business decision-making and management of daily operations. The key agent feeding information into the system is the field officer using the mobile application. Farmforce translates this into management information relevant for decision making in several ways: for example, combining the data collected due to the application of chemicals, the meta-information stored in the backend results in an internal control system that allows detection of irregularities such as the non-observation of application rules or waiting periods in real-time (*see* Fig. 6.2)

The management has not only an interest in production related information but also in monitoring and managing its field-staff in an efficient and effective way. To this end Farmforce provides views, not only on each farmer, but also on each field officer working in a scheme. Practitioners in development projects are aware that this level of transparency regarding field officers' performance can result in increased efficiency and effectiveness of their work provided that the management takes the opportunity to actively promote its staff performance. From agricultural and development perspectives, the data collected over consecutive growing cycles provide a rich source to observe changes in productivity over time as well as each farmer's profitability and relate this to any parameter related to production. Unsurprisingly, the capacity of Farmforce to track growing practice in detail also attracts users from research who are interested in studying crop management systems in developing countries.

6.5 Preliminary Results

Introduced in 2013, Farmforce is used in over 35 projects, managing around 150,000 farmers in 22 countries from Central America and the Caribbean to Western, Eastern and Southern Africa, and Asia, and a diverse range of 24 agricultural products as of early 2017 (Box 6.1). Since its commercial launch in 2013, Farmforce was successfully deployed not only in horticulture but also in other contract farming arrangements as diverse as cassava, coffee or cut-flower production. This fact proves the ability of Farmforce to cover very different contract farming arrangements with distinct documentation and certification requirements.

Both users and crops managed vary considerably which is according to the assumptions that have informed the development of Farmforce and is recognized as a strategy to professionalize smallholder farming (IFC, 2013; Kshetri, 2016; Protopop and Shanoyan, 2016).

For example, Doreo Partners, an impact investment firm in Nigeria, signed up to Farmforce in 2013 for its Babban Gona operations. At that time, Kola Masha, the social entrepreneur behind the venture, was running a maize intensification program with 2000 farmers spread over a total area of 1500 hectares in Northern Nigeria increasing their productivity by providing inputs and techn,ical advice, and buying their outputs. Kola subscribed to Farmforce to manage his operations and to be able to realize his ambitious growth plans which increased the number of farmers to over 12,000 by 2014 and is planned to reach 50,000 in 2020 (Storrs, 2014).

Similarly, Farmforce is used to manage the operations of Wilmar Agro Ltd, a small enterprise in Thika, Kenya, bringing to international markets Rainforest Alliance certified summer flowers produced exclusively by small and medium growers, with a view to improving their livelihoods (Atkins, 2013). Another enterprise using Farmforce and working with over 1500 smallholder farmers linked to a nucleus farm is Fair-Fruit in Guatemala, owned by the Dutch impact investment company Durabilis. Certified with FairTrade and GLOBALG.A.P. the primary market for their vegetables and fruits are the US market. With the introduction of the Food Safety Modernization Act requirements for documenting compliance with food safety standards have significantly increased driving demand for Farmforce in the Central American region.

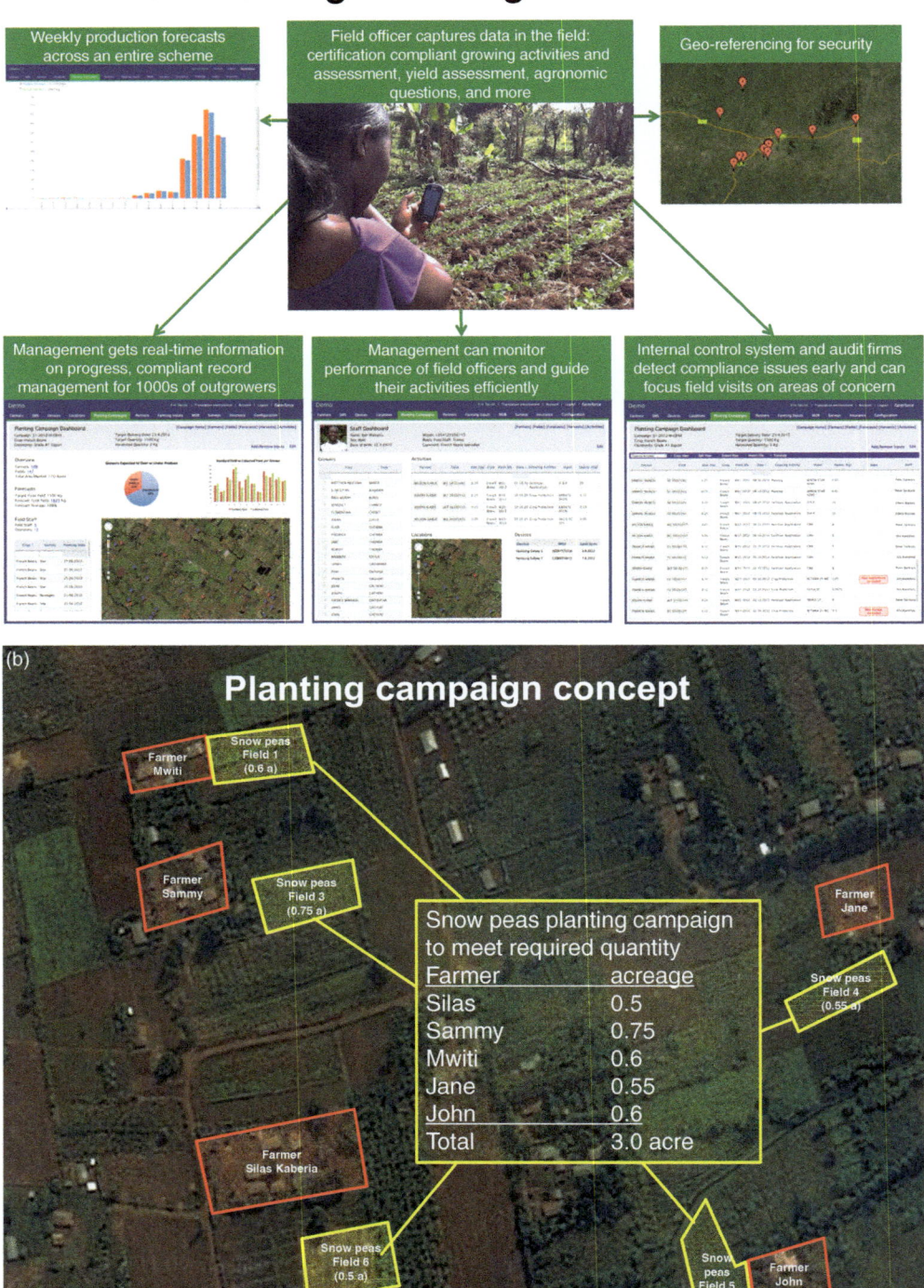

Fig. 6.2. Near real-time management information: (a) digital management; (b) planting campaign concept.

Box 6.1. Countries with Farmforce deployments (January 2017).

Africa: Cote d'Ivoire, Ghana, Kenya, Malawi, Mozambique, Nigeria, Rwanda, Tanzania, Uganda, Zambia
Asia: Bangladesh, India, Thailand, Vietnam, Singapore, Turkey, Indonesia
Latin America: Guatemala, Haiti, Honduras, Nicaragua, Peru
Crops managed using Farmforce: Baby corn, cassava, cereal, coffee, corn seed, French beans, garden flowers, groundnuts, maize, mangos, oregano, passion fruit, peanuts, peas, pepper, potatoes, rice, snow peas, soy, spice, sugarsnaps, tea, vanilla, vegetables.

A different – and growing – category of Farmforce users are big multinational companies such as Kellogg's, Tesco or McCormick who want to expand their engagement with smallholder farmers, often as part of their corporate responsibility strategy and sustainability commitments. McCormick, for example, started to use Farmforce as a means to facilitate procurement of sustainably grown oregano, black pepper and red pepper from smallholder farmers. Starting in 2016, the US-based company rolled out the system to about 500 farmers in Ivory Coast, 600 in India, 800 in Madagascar and 1200 in Turkey in the context of McCormick's participation in the Sustainable Spices Initiative. A wider rollout of Farmforce was slated for 2017, according to the company (McCormick, 2016).

Finally, development organizations started to use Farmforce as well. For example, the Clinton Foundation's Development Initiative (CDI) active in Rwanda, Malawi and Tanzania started to deploy Farmforce in 2014 to collect agronomic data from demonstration plots and smallholder farmers. Farmforce enables field officers to collect detailed information about growing and harvest activities, administer custom surveys, and facilitate market transactions. The insights enable the CDI to improve its programs and services. To this end, CDI tracks the output of their programs such as recording the number of trainings held, the amount of crops produced, or attendance at our community field days. Then, they analyze how this output has had an impact on people's lives measuring increases in productivity and profitability, access to markets and services (Clerkin, 2015; Kshetri, 2015).

To date, no systematic impact evaluation of the use of Farmforce has been conducted. However, anecdotal evidence documented by Farmforce users seems to confirm that Farmforce can reduce transaction cost in a number of ways:

- *Improved compliance and agricultural management:* one company notes that the time required for audit preparation went down from 15 to 3 days, pesticide detection went down by 53% and the rejection rate decreased from 10 to 2% following the deployment of Farmforce. Another company even reports zero MRL failure since Farmforce was deployed. In some instances, yield forecast accuracy has doubled which is particularly important given the market commitments of the aggregators.

- *Improved traceability:* farmer level traceability has become reliable according to many Farmforce users. In some instances, this has not only positive effects for downstream market integration but also increases internal efficiency and transparency. In one case – and people familiar with agriculture projects will not be surprised – ghost famers and famers that are not active any more were detected and could be eliminated leading to a reduction of 40% of farmers that are listed.

- *More effective use of field officers' time:* the mobile-technology-based data recording system has allowed field officers to spend 20 minutes more per visit to provide technical assistance to farmers – this time was used for manual report writing before and now contributes to increase farmers' know-how and productivity. Another Farmforce user reports that field staff effort has reduced by 30% and the real-time management enabled field staff to better meet defined key performance indicators (KPIs).

An interesting observation can be made across Farmforce deployments that, in itself, documents the effectiveness Farmforce introduces into contract farming management: moving from paper-based to IT-based management systems tends to evoke resistance once staff realize that the transparency that comes with Farmforce exposes them to a new level of oversight and accountability. In addition, Farmforce leads to a re-think and adaptation of the way operations are managed. All users confirm the importance of the engagement of the management during the introduction phase together with support provided for field officers to learn the new technology and understand the benefits of the system as key factors to render the Farmforce deployment successful.

6.6 Conclusions and Outlook

The introduction of Farmforce has advantages for aggregators and managers of contract farming arrangements at different levels, from more accurate and near real-time management information to strengthened compliance with VSS and better traceability and downstream integration.

However, the importance of tracking information on production, compliance and finance not only has value for commercial aggregators, it has wider relevance for the farmers as well: one of the biggest challenges small farmers face when they try to access finance or interact with formal players in the value chain is that they are an unknown risk. A big part of their economic life is informal and there is basically no formal record of what they are doing. Without the ability to make a credit assessment, banks and other financiers are reluctant to lend and other buyers cannot access the reliability of those farmers either. Records collected through Farmforce can give lenders and other formal market players confidence, even in the absence of traditional credit data. With the collection and sharing of records of input purchases or in-kind production credits, compliance with standard requirements and crop yields, farmers are finally gaining an economic and financial identity against which banks, micro-finance institutions (MFIs) and other commercial players are

increasingly willing to lend (Castell 2014). The 'economic identity' that Farmforce helps to establish itself is an important contribution to enable farmers' participation in formal markets beyond the initial contract farming arrangement.

Farmforce is designed as management software in the first place. However, the platform could expand in various directions to further support contract farming while sticking to its product approach. For example, the agricultural support structures built into Farmforce are robust but still relatively simple, providing:

- Active guidance to following a 'crop protocol' (i.e. the standard agricultural practice defined for a particular crop).
- An internal control mechanism for following application guidelines as described above.
- The possibility to send pictures of a disease stricken crop to an agronomist who sees the disease problem in the full context of the production.

However, agricultural support can be taken to a next level towards precision agriculture. Precision agriculture aims at site-specific crop management with the goal of optimizing returns on inputs while preserving resources (McBratney *et al.*, 2005). One such step Farmforce lends itself towards is linking the information on crops under cultivation contained in Farmforce with disease forecast models. Disease forecast models predict the likelihood of disease pressure which typically depends on the prevalence of microclimatic conditions conducive for a disease. The relevant parameters fostering diseases are typically temperature, rainfall, humidity and leaf wetness prevailing over a certain period of time. For example, Light Blight (LB) is a frequent disease in potato cultivation and LB models have been described as early as 1975 (Krause, 1975; Fry *et al.*, 1983). Knowing how disease pressure evolves over time allows prediction of the need for fungicide sprays early on. Early protective measures help to save cost on chemicals as well as reduce crop loss. With the advent of low-cost weather stations that transmit sensor data to cloud-based computing systems the automated monitoring and calculation of disease pressure for specific locations has become widespread in high-value agriculture such as horticulture, orchards, potatoes, tomatoes or coffee among others. While this holds for developed countries, farmers in developing countries lack access to this kind of information contributing to higher input costs and lower yields. Farmforce knows exactly 'who plants what' and opens new opportunities: low-cost weather stations put in farming areas monitor microclimatic conditions, send the data to a cloud-based computer which calculates pressure for a number of diseases. This information can be extracted and sent to Farmforce. Farmforce then selects all fields that currently have crops in a field that need preventative or early treatment against the disease identified and automatically sends instructions to the farmers and/or field officers in charge.

Finally, while the anecdotal evidence on Farmforce's effectiveness is encouraging, more rigorous research is required in order to understand the impact of Farmforce. The following questions could be part of a research agenda:

- *From an economic perspective:* what happens to the savings from reduced transaction costs? Are more farmers involved in outgrower schemes or do the savings translate into higher profits for the company or even into lower export prices that benefit consumers in the Global North?
- *From an agricultural development perspective*: how does the availability of proper record keeping and real-time agriculture information affect productivity and compliance with standards?
- *From an agricultural management perspective*: how does the transparency effect introduced by mobile technology influence management culture and staff qualifications in aggregator companies?

Acknowledgements

Farmforce was funded by the Syngenta Foundation for Sustainable Agriculture (SFSA) and the Swiss State Secretariat for Economic Affairs (SECO).

References

Abebe, G.K., Bijman, J., Kemp, R., Omta, O. and Tsegaye, A. (2013) Contract farming configuration: smallholders' preferences for contract design attributes. *Food Policy* 40, 14–24. DOI: 10.1016/j.foodpol.2013.01.002

Atkins, W. (2013) Professionalizing smallholder organization in Africa. In: This is Africa – A Global Perspective. *The Financial Times*, London, 28 May 2013. Available at: http://www.thisisafricaonline.com/Analysis/Professionalising-smallholder-organization-in-Africa?ct=true (accessed 28 February 2017).

Bellemare M.F. (2012) As you sow, so shall you reap: the welfare impacts of contract farming. *World Development* 40 (7), 1418–1434.

Bijman, J. (2008) Contract farming in developing countries – an overview. *Working Paper*, Wageningen University, Netherlands.

Brugger, F. (2011) Mobile applications in agriculture. Syngenta Foundation, Basel, Switzerland. Available at: https://pdfs.semanticscholar.org/c579/b4d800bb866eba834425cafef8204c5c3873.pdf (accessed September 2017).

Cagan, M. (2008) *Inspired: How to Create Products Customers Love*. SVPG Press, San Francisco, California, USA.

Castell, H. (2014) Innovative thinking helps close credit data gap. Available at: http://www.txfnews.com/News/Article/2945/Innovative-thinking-helps-close-credit-data-gap (accessed 28 February 2017).

Clerkin, T. (2015) How we're using data to empower smallholder farmers. Clintonfoundation.org blog, 23 September 2015. Available at: https://www.clintonfoundation.org/blog/2015/09/23/how-were-using-data-empower-smallholder-farmers (accessed 28 February 2017).

Collier, P. and Dercon, S. (2014) African agriculture in 50 years: smallholders in a rapidly changing world? *World Development* 63, 92–101.

Eaton, C. and Shepherd A.W. (2001) Contract farming: partnerships for growth. *FAO Agricultural Services Bulletin 145*, FAO, Rome.

Fry, W.E., Apple, A.E. and Bruhn, J.A. (1983) Evaluation of potato late blight forecasts modified to incorporate host resistance and fungicide weathering. *Phytopathology* 73, 1054–1059.

Fullbrook, D. (2014) Contract farming: what works and what might work better? Paper presented at the Contract Farming Workshop Conference, Vientiane, Laos, 4 February 2014. Available at: https://www.researchgate.net/publication/265824253 (accessed 28 February 2017).

Gereffi, G., Garcia-Johnson, R. and Sasser, E. (2001) The NGO–industrial complex. *Foreign Policy* 125 (13), 56–65.

GIZ (2008) *Contract Farming Handbook: a practical guide for linking small-scale producers and buyers through business model innovation.* Deutsche Gesellschaft für Internationale Zusammenarbeit, Eschborn, Germany.

IFC (2013) *Working with Smallholders: a Handbook for Firms Building Sustainable Supply Chains.* IFC Sustainable Business Advisory Service, Washington, DC.

Kariuki, I.M. (2014) Transition to certification schemes and implications for market access: Global GAP Perspectives in Kenya, *Agricultural Sciences* 5, 1100–1111.

Kshetri, N. (2015) Big data deployment in assessing the creditworthiness of low-income families and micro-enterprises in emerging economies: Platforms, methodologies and business models. Presented at 15th International Conference on Advances in ICT for Emerging Regions (ICTer), Colombo, 24–25 August 2015. Available at: http://ieeexplore.ieee.org/stamp/stamp.jsp?tp=&arnumber=7377655&isnumber=7377646 (accessed 28 February 2017).

Kshetri, N. (2016) *Big Data's Big Potential in Developing Economies: Impact on Agriculture, Health and Environmental Security.* CAB International, Wallingford, UK.

Krause, R.A., Masie, L.B. and Hyre, R.A. (1975) Blitecast: a computerized forecast of potato late blight. *Plant Disease Reporter* 59, 95–8.

Kuwornu, J.K.M. and Mustapha, S. (2013) Global GAP standard compliance and smallholder pineapple farmers' access to export markets: Implications for incomes. *Journal of Economics and Behavioral Studies* 5 (2), 69–81.

McBratney, A., Whelan, B., Ancev, T. and Bouma, J. (2005) Future directions of precision agriculture. *Precision Agriculture* 6 (1), 7–23.

McCormick (2016) A world of responsible solutions, corporate social responsibility. *Interim Report.* Available at: http://www.mccormickcorporation.com/~/media/28F6FD7966D4453D87A79C0D9D7DC2C8.ashx (accessed 28 February 2017).

Minot, N. (2007) Contract farming in developing countries: patterns, impact, and policy implications. *Case Study 6-3, Food Policy for Developing Countries: The Role of Government in the Global Food System*, Cornel University. Available at: https://cip.cornell.edu/DPubS?service=Repository&version=1.0&verb=Disseminate&handle=dns.gfs/1200428173&view=body&content-type=pdf_1# (accessed 9 May 2017).

Moore, G. (2014) *Crossing the Chasm*, 3rd edn. Harper Business, New York, USA.

Mumo, M. (2012) Horticulture exporters get tougher rules. *Daily Nation*, 9 February 2012. Available at: http://www.nation.co.ke/business/Horticulture-exporters-get-tougher-rules-/996-1323558-o66vj1/index.html (accessed 20 September 2016).

Mwambi, M., Oduol, J., Mshenga, P. and Saidi, M. (2016) Does contract farming improve smallholder income? The case of avocado farmers in Kenya. *Journal of Agribusiness in Developing and Emerging Economies*, 6 (1), 2–20.

Nguen, A.T., Dzator, J. and Nadolny, A. (2015) Does contract farming improve productivity and income of farmers? A review of theory and evidence. *The Journal of Developing Areas* 49 (6), 531–538.

Okello, J.J. (2015) Food standards, smallholder farmers and participation in high value fresh export markets. In: Hammoudi, A., Grazia, C., Surry, Y. and Traversac, J.B. (eds) *Food Safety, Market Organization, Trade and Development.* Springer International, Cham, Switzerland, pp. 205–227. Available at: https://link.springer.com/chapter/10.1007%2F978-3-319-15227-1_11 (accessed 9 May 2017).

Potts, J., Lynch, M., Wilkings, A., Huppé, G., Cunningham, M. and Voora, V. (2014) *The State of Sustainability Initiatives Review: standards and the green economy.* International Institute for Sustainable Development (IISD), Winnipeg, Canada and International Institute for Environment and Development (IIED), London.

Protopop, L. and Shanoyan, A. (2016) Big data and smallholder farmers: big data applications in the agri-food supply chain in developing countries. *International Food and Agribusiness Management Review, Special Issue* 19 (A), 173–190.

Rogers, E. (1983) *Diffusion of Innovations.* The Free Press, New York, USA.

Schireson, M. and Thakker, D. (2016) The money in open-source software. *TechCrunch (Crunch Network Blog)*. Available at:https://techcrunch.com/2016/02/09/the-money-in-open-source-software/?utm_source=ICTworks&utm_campaign=860b90f056-ICTworksEmailRSS&utm_medium=email&utm_term=0_0814c7961e-860b90f056-48308209 (accessed 20 September 2016).

Storrs, F. (2014) The solution to the global food crisis just might come from Nigeria. Available at: https://www.alumni.hbs.edu/stories/Pages/story-bulletin.aspx?num=3264 (accessed 28 February 2017).

Vodafone (2011) *Connected Agriculture: the Role of Mobile in Driving Efficiency and Sustainability in the Food and Agriculture Value Chain.* Vodafone, Berkshire, UK.

Wang, H.H., Wang, Y. and Delgado, M.S. (2014) The transition to modern agriculture: contract farming in developing economies. *American Journal of Agricultural Economics* 96, 1257–1271.

World Bank (2016) ICT in agriculture: connecting smallholders to knowledge, networks, and institutions. *e-Sourcebook Report no. 64605*, World Bank, Washington, DC.

7 Adoption of ICT Products and Services among Rice Farmers in the Northern Province of Sierra Leone

Simone Sala,[1]* Andrea Porro,[2,3] Alberto Lubatti,[1] and Stefano Bocchi[1]

[1]DISAA, Università degli Studi di Milano, Italy; [2]iMMAP, Washington DC, USA; [3]Faculty of Agriculture, University of Makeni, Sierra Leone

7.1 Introduction

Sierra Leone ranks close to the bottom (181 out of 188) in the Human Development Index (United Nations Development Programme, 2015). Almost half of its population (47%) lives on less than US$1.25 a day (World Bank, 2011), and the national status of food insecurity is alarming, rating 112th out of 118 in the *Global Hunger Index* designed by IFPRI (von Grebmer *et al.*, 2016). Agriculture is a key sector for Sierra Leone. It employs around 65% of the population, for a total of 400,000 family farms (FAO, 2016). Rice is the main food crop in Sierra Leone and rice farmers represent the backbone of the national food security system. It provides 40% of the overall caloric intake of Sierra Leonese (FAO, 2011) and it is a key crop to reach national food sovereignty (Conteh and Xiangbin, 2013). Compared to other crops, rice farmers worldwide need access to a much wider array of inputs and expertise, which suggests rice farmers can be early adopters of new technologies such as Information and Communication Technologies (ICTs). Focusing on rice farmers in Sierra Leone is thus particularly interesting at multiple levels.

In recent years, a variety of studies highlighted the positive role of ICTs in the agricultural sector (Chavula, 2014; Nakasone *et al.*, 2014). Nevertheless, most studies focus on the agricultural sector as a whole. Most global research rarely directly aims at understanding farmers' specific appropriation of ICTs and even less rarely takes into account the intra-seasonal behavioural dynamics that farmers experience, thus ignoring or downplaying that farmers need to take a variety of decisions during the agricultural production and post-production phases. As a result, the multi-faceted roles that farmers play as producers and consumers of information and knowledge across the agricultural value chain is scarcely acknowledged (Burrell and Oreglia, 2015).

International reports, press and blogs (among others) have been highlighting success stories about the positive role of ICTs in general, and particularly mobile phones, in the agricultural sector. Nevertheless, there are no large-scale studies about the diffusion of mobile phones among farmers in the Global South, nor evidence of the positive impact of mobile phones on farmers. Within this framework, the factors influencing mobile phone adoption and usage

---

* Corresponding author e-mail: salas@mit.edu

among farmers have been studied in different regions of the world (Islam and Grönlund, 2011; Martin and Abbott, 2011; Urassa, 2013; van Baardewijk, 2016).

The present chapter tries to contribute to the available knowledge on the impact of mobile technologies for agricultural development in Sierra Leone, aiming to show that the local context and networks play a strong role in determining the way rural communities adopt and utilize technologies in a West African country. Particularly, it investigates the way rice farmers in the north-eastern territory of Sierra Leone access different kinds of information across the agricultural season through a multiplicity of channels – including but not limited to mobile phones, as in the information model shown in Fig. 7.1. Furthermore, it seeks to identify the main demographic variables linked to adoption and use of mobile phones and the related services that can be accessed through these devices.

7.2 Data and Analysis — Sierra Leone

The research was carried out in the Bombali Shebora chiefdom and Port Loko district in northern Sierra Leone (see Fig. 7.2). The northern and eastern provinces are the most productive of Sierra Leone, also thanks to the highest availability of arable lands (Ministry of Agriculture, Forestry and Food Security in Sierra Leone, 2009).

Data collection was performed in a single stage in 2014. A group of 101 farmers was selected to represent both lowland and upland farming systems in the study area. Researchers from the University of Milan and University of Makeni (UNIMAK) trained enumerators to carry out the interviews with farmers in different local languages, so to be able to gain a representative picture of the Bombali Shebora chiefdom and Port Loko district. Farmers were profiled based on demographic attributes and their farming system, and they were interviewed to learn about their ownership and use of ICTs for agricultural production and marketing.

Table 7.1 shows the main demographic attributes of farmers in the sample. The sample is mainly composed of male farmers (around three quarters of the whole sample), with an average age between 40 and 49 years old. Unlike the situation in Europe (where the average age of farmers is 50), the different age groups are almost equally represented, thus suggesting that young people still engage in agriculture as it constitutes a viable source of livelihood. The large majority of sample farmers (65%) has an average income above US$640 per capita per year, with US$640 being the national reference bottom-line for poverty, thus confirming that rice farming can still be considered a remunerating activity. Only one house out of five has direct access to electrical power, with most houses located in the outskirts of urban centres that are

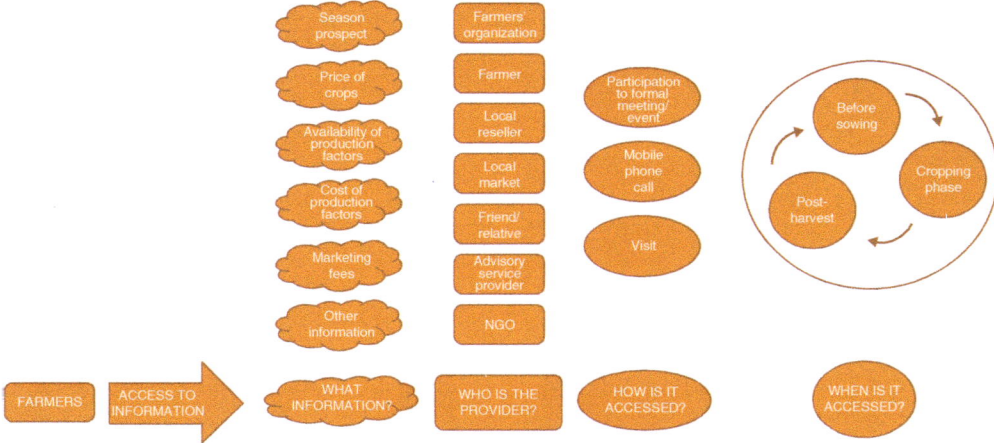

Fig. 7.1. Reference information model for farmers.

Fig. 7.2. Area of investigation.

Table 7.1. Demographic attributes of interviewed farmers.

Demographic attributes of farmers		(%)
Gender	Males	76
(*N*=93)	Females	24
Age	Below 39	25
(*N*=100)	40–49	39
	50–59	22
	Above 60	14
Average yearly	Below US$640	65
income	Above US$640	35
(*N*=97)		
Electrical power	Available	79
at home	Not available	21
(*N*=73)		
Education	No education	29
(*N*=73)	Primary school	28
	Secondary school	9
	High school	7
	Vocational school	12
	University	16

present in the area of investigation. Finally, two thirds of the interviewed farmers did not receive education or only accessed primary education, while the others received secondary-level education (9%), attended a high school (7%) or a vocational school (12%), or received a degree at the University (13%).

A PCA analysis helped identify 7 different clusters based on farm attributes such as farm size, farm management and crop system. These different clusters show specific peculiarities based on the farm attributes as well as the demographic attributes of the farmers themselves, as follows:

- Group 1 (*N*=61): common farms, close to the local average attributes.
- Group 2 (*N*=5): farms led by farmers with higher education level.
- Group 3 (*N*=8): farms led by young women, mostly farming only a single variety of rice.
- Group 4 (*N*=8): farms in Inland Valley Swamps, mostly composed by farmers with low-income practicing crop consociations.
- Group 5 (*N*=4): farms led by older women.
- Group 6 (*N*=6): farms led by older men, mainly in IVS who crop a combination of rice and cash crops.
- Group 7 (*N*=9): mixed farms.

With the aim of contributing to the available knowledge on the impact of mobile technologies for agricultural development in the Global South, the research tried to unveil two key aspects of the way farmers interact with ICTs through the interviews. First of all, the research analysed farmers' information networks and their use of mobile phones across the timeline of decisions they have to take during the agricultural season. Second, the study assessed the variables driving the farmers' adoption of such products and services.

Finally, farmers' profiles were cross-referenced with information related to their adoption and use of ICT products and services, which included: (i) availability of ICT products; (ii) access to ICT services; (iii) use of ICT products and services; and (iv) opinion regarding specific ICT products and services. Furthermore, farmers were asked about the way they look for information throughout the season and across different media channels, which highlighted a very dynamic behaviour of farmers across the season. Results of this analysis are presented in section 7.3.

7.3 Main Findings

7.3.1 Farmers' information, sources and channels

The tables below show the information that farmers look for across the production season (i.e., before sowing, during cropping and at post-harvest time) as well as their information providers. While it is understandable why farmers look for different information throughout the season, it is particularly interesting to highlight how the sources of information for farmers dynamically change.

Table 7.2 shows the information searched by farmers throughout the season. Much of the information searched by farmers is related to finance, and it includes: crop prices, cost of seeds, inputs, pesticides and transportation and intermediary fees. If grouped together, finance-related information represents the most important category of information for farmers during cropping (55%) and at post-harvest (96%). The fluctuation in the importance of finance-related information is particularly relevant as it differentiates the panorama of information consumption by farmers in the area of investigation from most existing studies and reports that tend to highlight the prices of crops as the pivotal

Table 7.2. Relative proportion of information searched by farmers.

Information searched before sowing	Proportion of information searched (%)
Season prospect	54
Price of crops	19
Price of seeds	15
Other	12
Information searched during cropping	
Labour cost	35
Availability of inputs	29
Cost of inputs	17
Availability of pesticides	10
Cost of pesticides	2
Other	6
Information searched after the harvest	
Crop price	67
Transportation cost	17
Intermediary fees	11
Other	4

information farmers look for at any time. Interestingly, seasonal prospects represent 54% of the information requested by farmers before sowing, thus highlighting an interest in crop outlooks. At this specific time of the season, only 34% of information is finance-related (i.e., 19% and 15% for the price of crops and seeds, respectively). For the rest, virtually the whole 'other' category includes different technical agricultural information (such as the weather, the right moment to see or transplant, soil fertility, etc.).

Table 7.3 shows the sources of information throughout the season. Farmers always represent more than half of information providers to farmers themselves: 52% before sowing, 60% during cropping and 48% at post-harvest time. At the same time, the relevance of extension and advisory service providers appears to decrease throughout the season: 15% before sowing, 7% during cropping and 2% at post-harvest time. Family and friends are relevant figures only during cropping, accounting for 5% of the sources of information (i.e., 50% of the 'other' category). Interestingly, trusted sources of information appear to be farmers themselves far more than

Table 7.3. Sources of information searched by farmers.

Information source	Information searched before sowing (%)	Information searched during cropping (%)	Information searched after the harvest (%)
Other farmers	52	60	48
Extension/advisory service providers	15	7	2
Local markets	12	11	34
Local resellers	10	7	10
NGOs	7	5	3
Other	4	10	3

third parties, including both institutional and private sector providers.

Farmers were also asked about their preferred communication channels (Table 7.4). In-person meetings with local farmers' organizations account for half of the preferences (51%), followed by mobile phone calls that total about one third of the preferences (29%). The mobile phone is the only non-interpersonal means of communication that appears to be used by farmers in the area of investigation. Indeed, the remaining 20% of preferred communication channels include in-person communication with other local stakeholders. Farmers pay visits to the local market or local acquaintances to access relevant information (7% in each of these two cases), or attend farmers' rallies when they are organized in the area (3%). Only one farmer referred to radio as a source of information, which was grouped in the 'other' category along with participating at meetings in the village and relying on an infomediary to access the information needed across the different phases of agricultural production and marketing.

7.3.2 Farmers' use of mobile phones

The large majority of the interviewed farmers (82%) have an active mobile subscription, in line with the national figure of 89% in the same year (World Bank, 2014). Distribution of mobile phones per family is high, as shown in Fig. 7.3, and only 5% of families do not have a mobile phone, while 3 out of 4 families have between 1 and 4 mobile phones.

Additionally, the data show that the number of phone calls made per day and received per

Table 7.4. Communication channels.

Communication channel used	Frequency of use (%)
Meetings with farmers' organizations	51
Mobile phone calls	29
Visits to local market	7
In-person visits to acquaintances	7
Farmer rallies	3
Other	3

day by farmers who own a mobile phone is relatively high: 60% of farmers make 0–5 calls per day and 40% make greater than 5 calls; 43% receive 0–5 calls per day and 57% greater than five.

Eighty five percent of farmers use their mobile phone for farming-related issues. Table 7.5 shows that the information most frequently searched relates to market prices (41%), advisory services (37%) or product buyers (22%). The table also shows that among the farmers who do not use a mobile phone, the majority declared that they would be mostly interested in advisory services (58%); 42% would look for market-related prices, and 8% would be interested to receive information from buyers.

To verify the consistency of such data, farmers were asked if they have the contact numbers of specific local agricultural stakeholders in their phone book (*see* Fig. 7.4). The data confirmed the availability of local contacts as most of them held telephone numbers of local market sellers (58%), while fewer of them have the contacts of agrochemical and input providers (23% and 21% respectively). Advisory

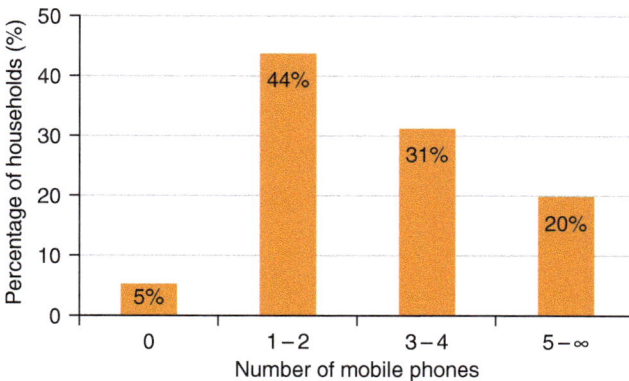

Fig. 7.3. Number of mobile phones per household.

Table 7.5. Information searched by mobile phone users and non-users.

Information source	Mobile phone users (%)	Mobile phone non-users (%)
Market price information	41	42
Advisory information (pests, innovations, etc)	37	58
Information from buyers	22	8

services are also well represented: extension agents and other advisory service providers are available in 30% of phone books, while local community organizations and NGOs account for 33% and 26% respectively. Interestingly, researchers' contact numbers are rarely available, confirming the low interaction with researchers expressed in Section 7.3.2. Indeed, the availability of phone numbers from the Sierra Leone Agricultural Research Institute (SLARI) and among UNIMAK staff is 13% and 3% respectively. These figures highlight the strong linkages between farmers and market stakeholders as well as local community organizations. Not surprisingly, given the low rate of input adoption, the contacts of input providers are limited.

The ownership of a smartphone is still limited among farmers, accounting for 17% in the overall sample. Among these, only one woman in the group declared owning a smartphone compared to 12 men. The share of farmers accessing the internet via mobile (29%) seems to confirm such data. Interestingly, Nokia is still the most popular brand, as 62% of mobile phone owners chose it. The Chinese company Tecno Mobile (selling Itel-branded mobile phones in Sierra Leone) ranks second, with 16% of preferences. This confirms the growing availability of Chinese phone brands in sub-Saharan Africa's mobile-phone market. Around one third of interviewed farmers (29%) can access the internet over their mobile, thus suggesting that at least one out of ten phones is a feature phone.

Interestingly, 68% of the farmers who do not have a mobile phone subscription use another person's mobile. The study thus confirms the key role of technology brokers in the rural settings of developing regions. The data show that these farmers mostly rely on family members (58%), friends (21%) or neighbours (11%) to access a mobile phone. Furthermore, these farmers mostly use third parties' mobile phones for business (50%) or for communicating with relatives and friends (41%). Concerning the use of mobiles for farming purposes, the data also show that farmers make 3 out of 4 calls to communicate with potential crop buyers, while the remaining phone calls aim at discussing with credit providers (17%) or input and agrochemical providers (8%).

A series of chi-square tests were performed to assess the significance of the relationship between the demographic attributes of farmers and their ownership of mobile phones, as well as their awareness of the services they could access through these devices. The typology of mobile phone (basic; feature; smart) and the availability of Internet access via the mobile phone was statistically correlated with farmers' income. Similarly, farmers' income was also linked with their awareness about the presence of a camera in

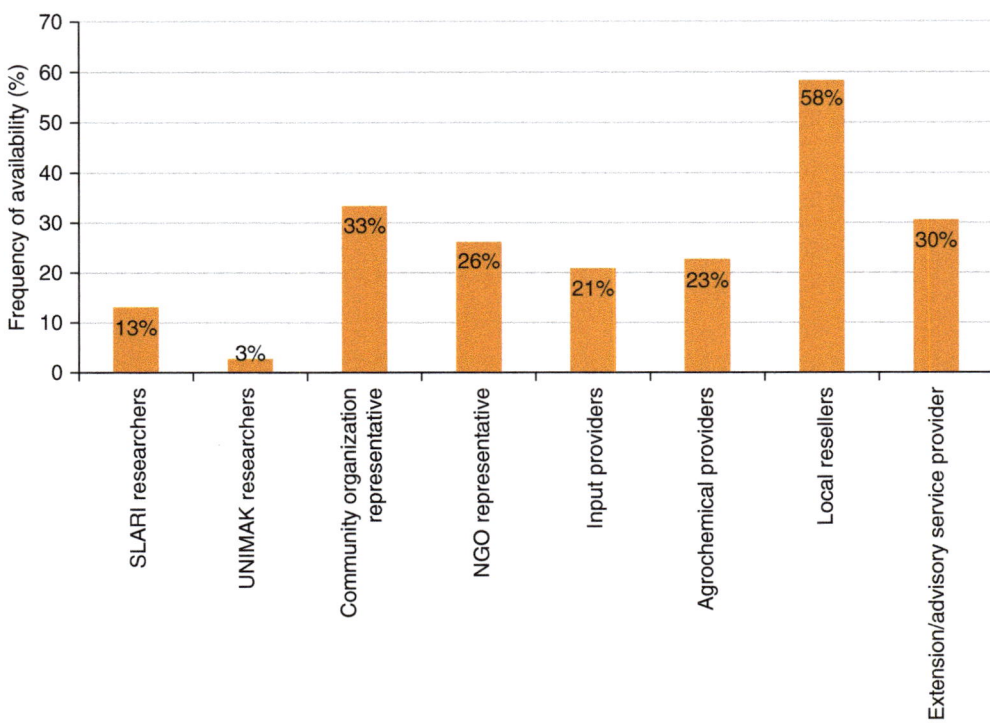

Fig. 7.4. Frequency of availability of local agricultural stakeholders' phone numbers in farmers' phone books.

their phones, as well as with the possibility to write messages in their local languages with their mobiles.

Furthermore, women more frequently installed the option to write in their local language and they tended to be more aware of the services offered on their mobiles (Table 7.6). A strong relationship was found between the typology of mobile phones owned by farmers and their ease of writing SMS and in English on their mobiles.

Table 7.6. Awareness of the possibility to write messages in local languages, by gender.

Awareness of the possibility to write SMS in local languages on owners' mobile	Males (n=63)	Females (n=14)
Yes	20	9
No	32	2
Does not know	11	3

7.4 Discussion and Conclusion

The chapter offers preliminary insights on the role of farmers as dynamic actors in rural knowledge ecosystems in the area of Bombali and Port Loko. Nonetheless, it should be highlighted that the results of the analysis do not allow generalizing the status of adoption and use of ICT products and services among farmers in Sierra Leone, mainly due to the size of the available sample.

In fact, the chapter presents the results of an exploratory study carried out before the 2014 Ebola outbreak that hit Sierra Leone, which did not allow follow up to the results with a larger-scale study. Such a study would have allowed assessing the significance of the relationship between the demographic attributes of farmers and their ownership of mobile phones, as well as the services accessed through these devices in relation to agriculture-related activities.

The results of the interviews challenged the usual portrayal of farmers as static consumers of information, showing that the kind of information that farmers look for and the related sources of such information do differ highly across the agricultural season. The study shows how finance-related information becomes important for farmers as the season unfolds, representing the most important category of information during production management (55%) and particularly at post-harvest time (96%). Peer-to-peer farmers' networks represent the primary source of information, and in-person meetings with other farmers or local stakeholders (e.g., resellers of agricultural products) are still the preferred way of sharing knowledge. At the same time, the study highlighted a high diffusion of mobile phones among rice farmers in the area of study, who use them to look for information by contacting market stakeholders as well as local community organizations. Access to the Internet also is already a reality for one third of the interviewed farmers.

While the availability of mobile phones is not linked with any demographic variable, income is clearly the variable that strongly influences the adoption of smartphones and the use of agricultural services among rice farmers of Bombali and Port Loko. It should also be noted that the study confirms the key role of infomediaries and technology brokers in the rural settings of the Global South, as almost 7 farmers out of 10 among those who do not own a

mobile-phone subscription reported the use of another person's mobile.

Despite the exploratory nature of the study, the results are important to inform policy-making supporting agricultural development in the area of investigation, with specific regard to rice producers. To effectively communicate with smallholder rice farmers it is important to take into account that 'information needs' rapidly change throughout the season, but other farmers remain the primary source of information at any time. It is thus pivotal to tap into inter-personal farmers' communication networks, even though mobile phones are now owned by a large majority of the rice farmers in the study area (82%) and thus represent a viable channel to communicate with smallholder rice farmers. Farmers recognize mobile phones as a valuable tool to look for information, and the integration of mobile phones with inter-personal communication methods available at the local level appears to be the best way to ensure effective communication with farmers in the area.

As a way forward, it would be interesting to conduct new studies in the country to confirm the validity of the trends captured with the present survey after the Ebola outbreak, by interviewing a larger sample of farmers. Furthermore, in order to generalize lessons on ICT for agriculture in Sierra Leone it would be recommendable to investigate the status of ICT access and use by other farmers' groups who are active on other cash crops, to verify adoption and use dynamics across smallholder farmers in the country.

References

Burrell, J. and Oreglia, E. (2015) The myth of market price information: mobile phones and the application of economic knowledge in ICTD. *Economy and Society* 44 (2), 271–292.

Chavula, H.K. (2014) The role of ICTs in agricultural production in Africa. *Journal of Development and Agricultural Economics* 6 (7), 279–289.

Conteh, A.M.H. and Xiangbin, Y. (2013) An assessment of the effect of price, policy and climate change ability on the supply of domestic rice in Sierra Leone: a supply response model approach. *International Proceedings of Chemical, Biological and Environmental Engineering* 60, 79–85.

FAO (2011) FAOSTAT: Food balance/food supply – crops primary equivalent. Available at: http://faostat3. fao.org/browse/FB/CC/E (accessed 1 October 2016).

FAO (2016) Sierra Leone at a glance. Available at: http://www.fao.org/sierra-leone/fao-in-sierra-leone/ sierra-leone-glance/en/ (accessed 1 October 2016).

Islam, M.S. and Grönlund, Å. (2011) Factors influencing the adoption of mobile phones among the farmers in Bangladesh: theories and practices. *International Journal on Advances in ICT for Emerging Regions* 4 (1), 4–14.

Martin, B.L. and Abbott, A. (2011) Mobile phones and rural livelihoods: diffusion, uses, and perceived impacts among farmers in rural Uganda. *Information Technologies and International Development* 7 (4), 17–34.

Ministry of Agriculture, Forestry and Food Security in Sierra Leone (2009) National Sustainable Agriculture Development Plan 2010–2030 (NSADP). Government of Sierra Leone, Freetown, Sierra Leone.

Nakasone, E., Torero, M. and Minten, B. (2014) The power of information: the ICT revolution in agricultural development. *Annual Review of Resource Economics* 6 (1), 533–550.

United Nations Development Programme (2015) *Human Development Report 2015*. Oxford University Press, New York, USA.

Urassa, N.S. (2013) *The use of cell phones in communication and dissemination of market information for beef cattle smallholders in Mpwapwa District, Tanzania*. MA dissertation, Sokoine University of Agriculture, Morogoro, Tanzania.

van Baardewijk, M.M. (2016) *Is the mobile phone the panacea for rural poverty? A case study on phone use among poor farmers in the Lucknow area, India*. MSc. thesis, Utrecht University, Netherlands.

von Grebmer, K., Bernstein, J., Nabarro, D., Prasai, N., Amin, S. *et al.* (2016) *2016 Global Hunger Index: Getting to Zero Hunger*. International Food Policy Research Institute, Washington DC.

World Bank (2011) Poverty headcount ratio at $1.25 a day (PPP) (percent of population). Available at: http://data.worldbank.org/indicator/SI.POV.DDAY/countries/SL?display=graph (accessed 1 October 2016).

World Bank (2014) Mobile cellular subscriptions. Available at: http://data.worldbank.org/indicator/IT.CEL.SETS.P2 (accessed 1 October 2016).

8 The Effect of ICTs on Agricultural Distribution Channels in Mexico

Luis Emilio Lastra-Gil*

Information Systems and Innovation Group, London School of Economics, UK

8.1 Introduction

Rural areas in developing countries are witnessing an impressive diffusion of Information and Communication Technologies (ICTs), which have transformed the traditional supply chain in a number of industries (Wiggins, 2014). Their impact on a sector where principles of family, kinship, and social connectivity have historically been especially powerful, is of great interest.

The potential benefits of ICTs include better access to markets, training and finance, easier building of collaborations, and more effective negotiation of deals. Medium-scale producers (working between 50 and 200 ha of land) are more likely to benefit from the use of ICTs (Walsham, 2010) and in Mexico comprise 2% of farms, working 16% of agricultural land (INEGI, 2007). They tend to be run by more affluent, better educated farmers who can afford both ICT infrastructure and to diversify their agricultural business. The two major areas of agricultural decision making are production and distribution as described in Fig. 8.1.

The focus of this chapter is on distribution channels, where ICT promises reduced transaction costs (and therefore optimisation) through facilitation of links between producers, markets and end consumers (Malone *et al.*, 1987). ICTs offer obvious communication advantages that will affect any exchange of goods or services,

and the mobile phone is a particularly seductive object of study, nearly ubiquitous alongside profoundly changing communication norms in people's daily lives, and generally assumed to have made agricultural distribution more efficient (De Silva and Ratnadiwakara, 2008). Nonetheless, the question remains exactly how ICTs help – or hinder – relationships between actors in the agricultural sector. Better communication and easier integration is but one element of rural development, an issue more complex than simple infrastructure alone.

This chapter is organized into six sections. In the next section the theoretical foundation of ICTs in agricultural distribution channels are reviewed, and the model and research questions are discussed. After a description of the research methodology and the case studies, the results and conclusions are presented.

8.2 Distribution Channels

A distribution channel is the chain of intermediaries through which a good or service passes until it reaches the end consumer. *Traditional intermediaries* are market actors that provide matching services for buyers and suppliers in a market, the location of goods and services, and their characteristics (Chircu and Kauffman, 1999). These roles can be carried out by trade

* E-mail: L.E.Lastra-gil@lse.ac.uk

© CAB International 2018. *Digital Technologies for Agricultural and Rural Development in the Global South* (ed. R. Duncombe)

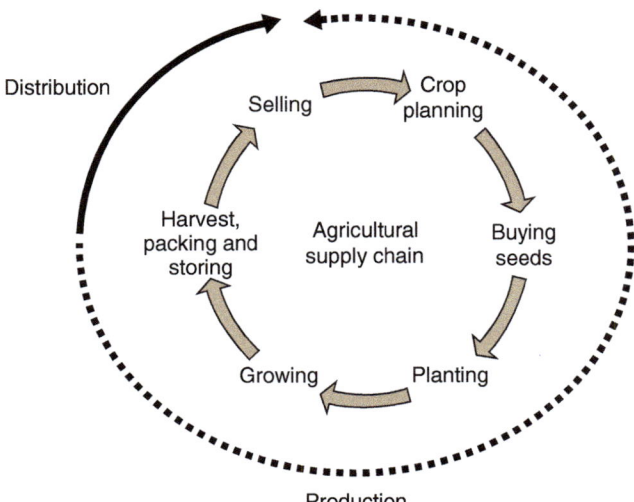

Distribution

Selling — Crop planning

Harvest, packing and storing — Agricultural supply chain — Buying seeds

Growing — Planting

Production

Fig. 8.1. Agricultural business decision process.

fair, by a broker, distributor, wholesaler or retailer (Cateora *et al.*, 2011). *Non-traditional intermediaries* are relationships supported by ICT, for example, the internet or electronic marketplaces (Chircu and Kauffman, 1999), as illustrated in Fig. 8.2.

Traditional intermediaries include:

- *Trade fairs* (or trade shows or expos) are B2B exhibitions where firms showcase and demonstrate products and services, meet with current or prospective industry partners and customers, contact and evaluate potential agents and distributors, study rivals, and examine market trends and opportunities (Cateora *et al.*, 2011).

- *Brokers* are individuals or parties that arrange transactions between a buyer and seller for a commission when the deal is executed. They act on behalf of the producer but do not take ownership of products and services. In Mexico, brokers tend not to prepare formal contracts and are often given the pejorative label 'coyote' (Keys, 2005).

- *Distributors* are commercial entities that buy products and warehouse them, and resell to retailers or direct to end-customers. Most distributors provide financial support to the supplier. They often also offer services including product information, quotes, and after-sales technical support (Cateora *et al.*, 2011).

- *Wholesale* is the resale or sale without transformation of new and used goods to retailers, institutional or professional users or other wholesalers, e.g., the public wholesale markets located in every large city in Mexico, also called 'Centrales de Abastos' (Schwentesius and Gómez, 2002).

- *Retailers* are economic actors that sell consumer goods and services to end consumers through multiple channels of distribution, from mobile street markets or 'tianguis' that change location from day-to-day (Schwentesius and Gómez, 2002) to supermarkets that increasingly offer online services and home delivery.

Non-traditional intermediaries favour ICT-mediated relationships which can improve transaction efficiency because firms can disperse business among many competitors, sampling prices widely, avoiding the use of small-number bargaining and entrapment. However, despite their increasing popularity and obvious success in other industries, doubts persist over the suitability of electronic marketplaces for the agricultural sector. Few are to be found outside of cattle and dairy sub-sectors, and it is unsurprising that empirical research on them in developing countries is scarce (Brush and McIntosh, 2010; Cloete and Doens, 2008).

Seitz (2013) describes several factors that have affected the development of business-to-business (B2B) e-commerce in the agriculture

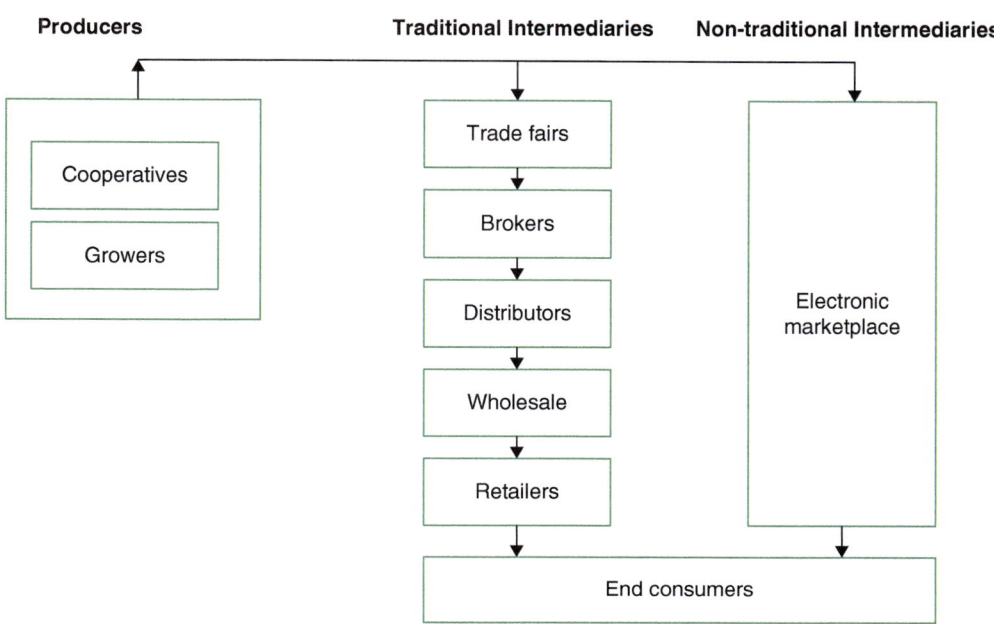

Fig. 8.2. Agricultural sector distribution channels.

sector. The supply chain in agricultural transactions is complex, and there have been changes in industry structure (principally concentration of large agribusiness and retail) alongside shifts in agricultural production and consumer preferences. Brush and McIntosh (2010) found that scale of production is an important factor in adopting e-commerce. Infrastructure limitations (particularly for smaller-scale producers) include financial barriers to reliable internet access, poor user proficiency, and security risks to both suppliers and purchasers.

In agricultural e-marketplaces, transactions occur over virtual platforms for bartering, commodities trading and auction. While many e-commerce platforms remain at the proof-of-concept or scalability stage, few have achieved sustainability, and have tended to be concentrated in non-Latin markets. For example, Google is linking buyers and sellers through mobile and internet-based platforms in Africa (Baumüller, 2012), and in China the e-commerce company Alibaba.com has become a vital agricultural business information platform, a typical B2B model that provides services for small and medium enterprises across the agricultural ecosystem (Bao *et al.*, 2012; Yanyan, 2015).

8.3 Current Theoretical Perspectives

Two different perspectives have been applied to the study of ICTs in the type of decision making that is common in the agricultural sector (*see* Fig. 8.3). Institutional Economics emphasizes the importance of information and communication in coordinating and simplifying economic transactions via contracts, organizational structures, language, culture and human behaviour (Coase, 1937; Williamson, 1985), while New Economic Sociology argues that individual choices and actions are instrumental but situationally constrained by the networks in which individuals are embedded (Granovetter, 1985).

8.3.1 Transaction Cost Theory

Using an interdisciplinary approach from economic and organization theories, Transaction Cost Theory attempts to identify the institutional form that provides the most efficient exchange under conditions of individualism, utility maximisation, bounded rationality and opportunism. Transactions – the transfer of property rights between at least two parties – are the fundamental unit of human economic interaction (Williamson, 1985). They are analysed with respect to three attributes: asset specificity, uncertainty, and frequency. Asset specificity refers to the dependence created through transaction-specific investments (Collin and Larsson, 1993). Value arises either from the investment made by the parties in the exchange, or through the cost incurred by terminating the relationship for another exchange party. Uncertainty is linked to situations in which bounded rationality makes humans incapable of predicting the future, and frequency of transactions influences reputation. Transaction Cost Theory predicts that the 'market' will govern transactions characterized by a low level of transaction-specific investment, whereas 'hierarchies' preside for high transaction-specific investments. Developing this simple bipolar institutional framework, Williamson (1991) also introduced the 'hybrid' as an intermediate form between market and hierarchy, involving reciprocal trading, regulation, franchising and various forms of long-term contracting.

Our understanding of the influence of ICTs on governance structures has evolved over the years. In the earlier literature, markets with arm's length transactions were seen as the preferred governance structure. *Brokerage* (a *'move to the market'*) was predicted through a reduction in transaction costs and integration of adjacent steps in the value-added chain in electronic marketplaces (Malone *et al.*, 1987). Firms would outsource more and transactions would be conducted through open markets. However, this overlooked the tendency of firms to favour hierarchical relationships with fewer suppliers that more reliably result in the right product arriving at the right time, and which better protect against opportunistic behaviour (Koch and Schultze, 2011).

More recent work has decoupled the effects of ICT on transaction costs and exchange risk, proposing instead an *integration* (a *'move to the middle'*). This argues that ICTs and cooperative relationships are mutually reinforcing: firms will tend to choose hybrid modes of coordination that mix market and hierarchy logic. These 'mixed mode' governance structures characteristically take the form of long-term cooperative agreements with a few suppliers

(Grover *et al.*, 2002; Koch and Schultze, 2011). Many transactions remain in the middle between established markets and partnerships. Kambil *et al.* (1999) has noted that buyers and sellers face unexpected changes in demand, which demands a variety of transaction designs: the lowest cost might be either market spot price or a negotiated price with a trusted supplier. Buyers are motivated to move away from tightly coupled, or vertically integrated relationships; organizations towards a more market-like structure to reduce sourcing risks, and sellers can create margin opportunities by abandoning purely open markets.

Despite being home to many transactions that fall between pure markets and hierarchies, the 'middle' is considered to have high structural and operational uncertainty. *All-in-one-markets* have been proposed as a solution, a technology-enabled platform that aggregates multiple transaction modalities with market-oriented functionalities. For example, quote handling, cataloguing, supply chain management, and customized marketplaces; capabilities that facilitate electronic brokerage and integration (Kambil *et al.*, 1999; Koch and Schultze, 2011). This potentially leads to 'disintermediation', the displacement or elimination of market intermediaries through direct trade between producers and consumers. Electronic marketplaces potentially suppress intermediaries by allowing buyers to search directly for appropriate suppliers. However, an ICT-heavy ecosystem still requires coordination and collaboration; indeed, there exists the potential to increase transaction costs by un-gating an overwhelming volume of information (Cordella, 2009). An overabundance of information might increase the need for specialized intermediaries that can match customers and suppliers by aggregation, filtering, trust provision, and facilitation: thus the potential for resilience of intermediation (Bakos, 1998, Sarkar *et al.*, 1998), through realignment and re-intermediation of existing, or entry of new, actors.

Thus, Transaction Cost Theory predicts that ICTs should improve access to information, potentially reducing search, negotiation and enforcement costs. By facilitating the exchange of information, goods, services and payments associated with transactions, ICTs should better match buyers and sellers and better provide the legal and regulatory frameworks that enable efficient institutional function. While ICTs have subverted much traditional intermediation, they have not delivered what Transaction Cost Theory has promised, which is to eradicate asymmetry of information, and a more dangerous weakness of Transaction Cost Theory is its neglect of the network that surrounds (particularly agricultural) economic exchange.

8.3.2 Embeddedness Theory

Embeddedness Theory aims to explain innovation and the diffusion and adoption of technology by individuals and communities within a 'network', a socially embedded context that influences the uptake and use of ICTs and is therefore fundamental to understand socioeconomic interaction (Granovetter, 1985). The network can be defined by the geographical reach of the people who interact with each other (Molony, 2006), or as the regular contacts and social connections among individuals or groups (Granovetter and Swedberg, 2011).

Social networking especially has become the basis of a new type of networked communication. The study of social media is important because 'making the web social' has turned into 'making sociality technical' via platforms that manage and engineer daily human interactions. The back end of social media is far from silent: activities are catalogued, processed and sold to the paying customers (Van Dijck, 2013). Information is utilized in various ways, for example to personalize advertisements, as big data output, and as a contribution to the systematic personal profiling of data brokers. It is this new unregulated territory of information sharing that is called the 'culture of connectivity'. ICTs help exploit external knowledge outside the firm's boundaries for collaboration and manage the relationship (Van Dijck, 2013, Koch *et al.*, 2013). Farmers can, potentially, use social media to achieve a constant inflow and outflow of information which results in better learning opportunities.

Interactions in a socially embedded network are governed by the expectations and norms associated with that web of social attachments. Norms are part of the social capital in any social setting, which in embedded relationships tends to take the form of trust, goodwill, obligation and reciprocity. Social capital is an aspect of economic activity that essentially amalgamates 'who you know', what ordinary language calls 'connections', and can be exchanged for other capital, human and economic. An actor in an embedded relationship may exploit a sense of obligation or loyalty for more favourable treatment (e.g., discounts or expedited service) (Schultze and Orlikowski, 2004).

The literature argues that it is not price, but the exchange of supply and demand information between farmers and intermediaries where ICT artefacts should hold the greatest benefit, because it allows first-hand exchange of up-to-date information (Molony, 2006). Market prices are often irrelevant or subordinate to other factors for trade-related decisions (Burrell and Oreglia, 2015). Economists tend to present an abstracted view of the role of information in the market, removing prices from the trade practices and relationships in which they are embedded. Such relationships appear to be especially critical on smaller family farms. Existing business relationships, trust, attitudes towards risk and institutions, rules and policies are all inputs for decisions on what to produce and whether and to whom the farmer should sell.

Molony (2009) has argued that ICTs do not significantly alter trust relationships between agricultural producers and wholesale buyers. Farmers often lack credit to purchase agricultural inputs, and may rely on their buyers to provide it. They must accept a price irrespective of the technology used to communicate this information because their buyers are also their creditors. Farmers are unable to exploit ICTs for better information on market prices or to find new buyers because they risk breaking the long-term relationship that provides credit. Should ICT marketplaces connect farmers to the credit they require, it could permit escape from such exploitative buying. A significant weakness of embeddedness theories is their failure to consider the farmer's entrepreneurial and negotiating capacity to identify distribution channels and agree better deals.

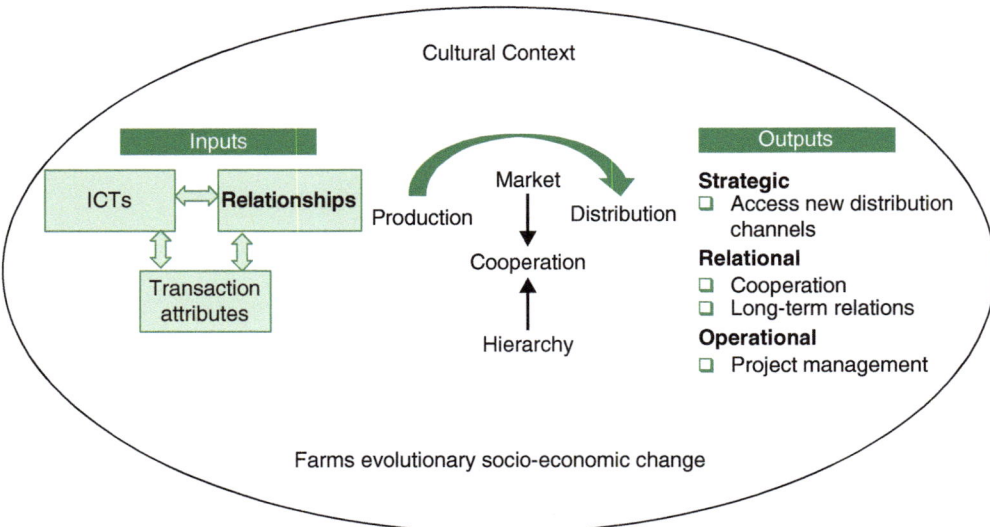

Fig. 8.3. Theoretical framework.

8.4 Methodology

The research poses the question: what role do ICTs play in the distribution of agricultural produce in a developing country? The research adopts an interpretive case study, using a qualitative approach with careful consideration of contextual conditions. This is an appropriate design to understand the governance structure in the supply chain network and the role that ICTs play in the distribution of agricultural produce. A theoretical framework developed through multiple iterations of data collection and analysis helped generate rich and empirically grounded insights. The research also captures everyday circumstances and conditions that should better inform about typical experiences of people and institutions, and affords the opportunity to observe and analyse a phenomenon in a specific context. Fieldwork was undertaken in Mexico in four phases between 2013 and 2016 (Table 8.1).

The selected case studies (Tabasco and Campeche) cover regions where mobile communication is more reliable and there is robust internet access: growers here were much more likely to use ICTs in their decision-making processes. Entry to the case studies was gained through a 'gatekeeper' in each region, an actor with control over key sources and avenues of

opportunity (Hammersley and Atkinson, 2007), who provided an introduction to other growers in his/her network. For example, the gatekeeper at Mexico City's wholesale market provided an introduction to a firm that grows and distributes bananas, ultimately leading to contact with the plantation owner in Teapa (Tabasco), who in turn provided an introduction to the other farms in the case study cluster. The head of the NGO 'Ayuda para Ayudar' (Help for Help), facilitated access to Maya farmers in Yucatan, and was the gatekeeper to the second case study farm in Palizada (Campeche). Data was collected from multiple sites for each case, and communication with the main actors continued via e-mail, instant messaging, and phone calls.

Semi-structured interviews were conducted using a funnel technique, beginning with context and wider issues and reaching details on the usage of ICTs for agricultural decision making. Purposive sampling offered regular contact in informal settings, building trust and facilitating access to new contacts. A total of 128 interviews were conducted (14 with gatekeepers, 42 with smallholders, 17 with medium-scale producers, 17 with wholesale merchants and 38 with community members that included local government officials, rural internet operators, and NGO staff). Observations were both observer-as-participant (outsider) and complete

Table 8.1. Research design and time plan.

Semi-structured interviews	Dec 2013– Jan 2014	Mar–Apr 2015	Jul–Aug 2015	Aug–Sep 2016	Total
Gatekeepers	5	7	3	2	17
Smallholders	5		38		43
Medium-scale producers		2	11	12	25
Merchants		14	3		17
Other key informants		2	23	15	40
Total	10	25	78	29	142
Re-interviews					
Gatekeepers			2	1	3
Smallholders				1	1
Medium-scale producers			2	6	8
Community members			1	1	2
Total Re-interviewed	0	0	5	9	14

observer (insider). Archival analysis provided statistical data, and thematic analysis was used to structure and process information from text sources as basic, organizing and global themes.

8.5 Case Studies

Two producers were selected for case studies, a cluster of banana growers from Tabasco and an organic rice producer from Campeche seeking new markets abroad. A summary of interviewees in provided in the Appendix. The Tabasco cluster is heterogenous and market-driven, but there is a relational element in the decision-making process (i.e., farmers buy agricultural inputs locally), while the Campeche cluster is a family-run estate and hierarchy-driven. The choice of different geographical areas furthers the generalizability of any findings (Walsham, 1995) and better represents the national picture of ICT use in rural areas.

8.5.1 Case A 'cluster': banana growers, Teapa

Mexico is the world's twelfth largest producer of bananas (SIAP, 2015). Teapa lies 60 kilometres south of Tabasco state capital Villahermosa and is home to a cluster of medium-sized banana plantations that apply collective expertise to produce a traditional crop with modern techniques. Tabasco has a geographic competitive advantage in this industry because it takes just three days to ship produce to the east coast of the US via the local ports of Frontera and Dos Bocas or the more distant but larger ports of Coatzacoalcos and Veracruz. The supply chain network in this cluster depends on domestic versus export destination.

Seventy producers, working a total of 10,000 ha, are registered with the '*Union Platanera*'[1], supported by a network of small and medium enterprises that sell agricultural inputs, airplane services, plastic bags, pallets, etc. The Banana Association seeks to improve productivity and provide export documents for carrier companies. Its remit extends to indirect support for banana producers, for example lobbying for better transport links. The association periodically audits producers, all of which have their own brands (for example Alta, El Refugio, Tony Bananas, Santa Rita).

High asset specificity (land specificity with a permanent crop, production process, time specificity to reach the customer, technical and human knowledge) and significant environmental and seasonal uncertainties in the banana sector have encouraged market-dominated transactions. For example, prices plummet with the production glut between summer and autumn with supply exceeding domestic demand and causing temporary debt for growers. In this respect, ICTs can help by reducing uncertainty and enhancing trust and reputation, by more efficient economic organization in terms of coordination costs. These coordination costs take into account the costs of gathering information, negotiating contracts and protecting against opportunistic behaviour. In the market, ICTs can help farmers to compare different buyers and select the one that provides the highest price and good terms and conditions. The market coordination costs are relatively high, because farmers must analyse information from a variety of sources in a short period of time with the pressure of selling fruit on a weekly basis, thus, opportunistic behaviour can arise. On the other hand, hierarchies restrict the choice of buyers to one predetermined customer, which allows better planning and control. These transactions reduce coordination costs by eliminating the need to analyse a great deal of information. The hybrid mode, encourages longer-term relationships with customers which encourages selection of distribution channels that involve longer-term contracts, providing security of revenue.

Table 8.2 describes the distribution channels for Teapa's banana growers, all traditional intermediaries. The first three scenarios are for the domestic market. In scenario one, there is direct negotiation with a wholesale market that controls end-to-end distribution; direct sale to the wholesale market is the most important distribution channel. Terms and conditions are typically 1–2 weeks' credit. In the low season at least, there is no reliance on brokers (scenario two) unless there is an urgent need to sell fruit quickly, avoided by medium-scale growers despite paying in cash because their commission reduces margins.

The wholesaler adds value to the fruit. They have a ripening chamber to make the fruit yellow. They can keep it green for up to 40 days depending on wholesale commitments.

(A2, see Appendix)

They [brokers] have connections and tacit knowledge of the market while growers have production know-how.

(A2)

In scenario three, producers add value to the fruit by ripening it themselves and selling direct to a retailer, a disintermediation of the wholesaler. Supermarkets offer better prices and longer terms, determining spot prices weekly, but with no annual contracts. As registered suppliers, producers must provide a quotation each week via e-mail to maintain their business relationship.

Supermarkets have a stronger market share in the volume of fruits traded in Mexico. As producers, we must be conscious of trends, open-minded and change traditions. It's an option, but not the solution for the banana distribution channel.

(A1)

Scenarios four and five describe distribution channels for the export market. Growers sell to export distributors, and large agribusiness concerns, directly or via brokers. Often meeting at international fairs, potential export buyers will first visit the Banana Association. Face-to-face meetings are important and have become the norm over 30 years of negotiating export contracts. Relationships are subsequently maintained with e-mails and phone calls. Initially the buyer carries risk by paying in advance; transactions are typically weekly and become more flexible after one year. Now common practice, this has helped establish trust and reputation in the distribution channels, a relational factor reducing transaction costs mediated by ICTs.

Table 8.2. Distribution channels for a banana cluster.

Scenario	Trade Fair	Broker	Wholesale	Internet	Distributor	Retailer
1. Wholesale			X			X
2. Broker		X	X			X
3. Supermarket						X
4. Export Broker	X	X			X	X
5. Export Distributor					X	X

Exports are based on an annual contract, and are very useful when there is overproduction for the domestic market. Exports help to mitigate exposure to declines in domestic price due to fluctuations in supply, principally from seasonality.

(A2)

I sell 70 percent of production to a Spanish customer. Perhaps I am less averse to market risk and price fluctuations than my colleagues, but I avoid speculation in the domestic market. The security of a long-term contract allows me to plan working capital and business needs without the complexity of dealing with several distribution channels. This also helps me to plan capital investment on my ranch.

(A3)

If an export customer calls me now and needs a container [1300 boxes] I make the arrangement and send it. If I am unable to complete the whole purchase order alone I work with other producers to complete the shipment.

(A3)

In 2016, twelve banana producers contracted to supply Chiquita Banana, an international agribusiness corporation, under 'free carrier' (FCA) terms:

The contract with Chiquita Brands is FCA, we are responsible for production and harvesting components and labour for packing. They are responsible for land and maritime transport from the ranch to the port and from the port to the final destination.

(A2)

A longer-term contract with large agro-corporations reduces price volatility, hence cooperation among members of their community will allow them to reach larger markets and achieve economies of scale.

The banana export market has been transformed by a free market that has brought flexibility in export and transportation. Banana producers aim to sell at least one third of production to the export market through contract farming. The need to satisfy a challenging market has encouraged collaboration between suppliers, producers and customers. The importance given to establishing longer-term term contracts with commercial partners points to efficiently mediated modes of governance in agricultural markets.

Our goal is to match banana supply with international and especially European demand at the end of the year when the domestic market is saturated. We not only help the national economy, but also those producers who are primarily focused on the domestic market.

(A3)

ICTs are used as a project management tool for planning and control of production, intra-cluster cooperation, negotiation, and internal and external communication. Business meetings with customers are carried out face-to-face; minutes and formal communication are documented and distributed by e-mail, but daily informal communication (voice, text messages, photos, video) is via the free messaging platform WhatsApp where producers have created groups for communication. Farmers that export are more conscious of the advantage that the internet offers.

In Mexico, there is no e-marketplace where buyers can place a small purchase order [for bananas] (e.g. 100 boxes). We don't deal with small purchase orders because we need to send an entire container, possibly a wholesaler can do this. Here the factory, producers and buyers interact to agree terms and conditions. We monitor purchase orders, payments and so on with mobile phones, emails, text messages, WhatsApp. In the past, we used fixed voice, telegrams or the post.

(A2)

I am primarily focused on the export market. The internet is very effective for finding new markets and suppliers and to make yourself known. I have added several videos on YouTube to promote my company, and have a Facebook presence[2], but voice telephony helps me to negotiate and close the deal.

(A3)

For domestic transactions, the mobile phone is an important tool and has facilitated and streamlined the processes of communication, getting closer to customers and suppliers. It has helped us to make negotiations more efficiently and faster.

(A5)

The banana sector is quite personal, encouraging direct contact and face-to-face interactions. This can explain why farmers have not fully exploited ICTs and found new distribution channels. ICTs and cooperative relations have enabled better collaboration and communication among growers and buyers. ICTs have facilitated the change from an entirely market-based governance to a hybrid mode of economic organization that mixes markets and hierarchy. Clustering enables deals with multinational corporations, a major factor supporting the embeddedness view of the economy.

8.5.2 Case B 'family single estate': Organicos del Tropico, organic rice grower, Palizada

The town of Palizada once had fertile land, with 60,000 ha of woodland cleared for cattle and rice production in the 1970s. However, inappropriate use of pesticides and fertilizers has exhausted much of the land; and only 5000 ha remain productive today and the infrastructure created to support rice production has been largely abandoned. The once wealthy local association of rice growers has gone. Organicos del Tropico at Rancho Pancho Villa began operating in the early 2000s using an alternative organic approach for rice growing. Its proprietor hopes to make the land more fertile using natural products rather than industrial chemicals. Fertility has been improved by buffalo grazing. This case offers an example where ICTs and especially the internet have been used successfully to expand an agriculture business.

As for *Case Study A*, the attributes of organic rice transactions have influenced selection of efficient hierarchical governance. End-customers tend to purchase these products in specialized shops or high-end supermarkets. Farmers deal directly with the purchase departments of these distribution channels. Transactions are predominantly with high asset specificity in land, specific organic production process, technical and human knowledge, and organic registration. Distribution channels also require certified organic warehouses. Organic rice producers aim to reduce asset specificity, uncertainty and increase the frequency of transactions, encouraged by a move to the middle through a shift to electronic market transactions and longer-term relationships, which is in the opposite direction, but symmetric to what was observed in *Case Study A*. The following table presents the distribution channels for Palizada's organic rice growers. Table 8.3 presents the distribution channels for Palizada's organic rice grower.

Selling organic produce is difficult, requiring certified warehouses. Scenario one describes the relationships with domestic supermarkets via distributors selling their and the farm's own 'Pijije' brand[3]. Distributors usually buy a variety of products from different suppliers and this generates economies of scale. Scenario two also invokes traditional intermediaries and describes courting of specialized organic retailers at trade fairs, mostly independent shops on the Yucatan Peninsula, for whom buying directly from a farm is more convenient and cheaper than purchasing through a distributor in Mexico City.

Scenarios three and four are non-traditional and demonstrate how social media and a webpage respectively have enabled this producer to reach new customers domestically and abroad. Approximately 10% of its customer base has been gained through Facebook, permitting escape from traditional distribution channels through partial disintermediation. The grower is a member of a group where people share ideas about organic rice cultivation and has contacts worldwide, including in Australia, Brazil, Italy and Colombia.

> Facebook is a marketplace and a useful platform for communication, a good source of information that offers guidance on what customers like and dislike.
>
> (B1)

Table 8.3. Distribution channels for Organicos del Tropico.

Scenario	Trade Fair	Broker	Wholesale	Internet	Distributor	Retailer
1. Domestic supermarkets					X	X
2. Domestic organic retailers	X					X
3. Social Media				X		X
4. Company webpage				X	X	

People who consume organic food are middle or upper-middle class who usually are able to communicate and provide comments. Consumers give me new ideas and show how the market is moving.

(B1)

Customers place orders through the firm's Facebook page. The main challenge is delivery logistics. Costs are not standardized across Mexico and with the rice plantation 200 km distant from the delivery company's hub, purchase orders are consolidated fortnightly to save costs. The grower has also achieved disintermediation through the internet:

While it was not a goal to reach end consumers directly, we are building this distribution channel. It is an alternative way to increase revenue. We began with a small number of customers, but there is a potential for growth. Because distributors no longer provide this service, I thought that I should do it myself with the help of the internet.

(B1)

The first question we ask when potential customers contact us is their city and state. Even small businesses from Peru, Guatemala, Costa Rica have contacted us. I cannot support them because these are outside of Mexico and logistics demands economies of scale. They sometimes ask for only a box. The cost of the shipment would exceed the cost of the product.

(B1)

For many years, the grower sought a customer prepared to invest as a business partner, to support rather than abandon, should for example product demand collapse. He is currently engaged in developing a product with German firm HiPP GmbH & Co. Vertrieb KG, which approached him via his website.

We are developing twelve different varieties of Japonica rice. We started with 100 grams, followed by 15 kilograms and then 300 kilograms of each variety. I have allocated 20 ha for their product.

(B1)

If I tenant my father's [currently uncultivated] land I can supply a big client. But I need a good relationship for security; I would like them to invest. They [HiPP GmbH] have been involved in discussions and they have said yes. The company has a representative in Costa Rica.

(B1)

The grower alone has been responsible for business development, as profits are too small to permit hiring a specialist. ICTs have helped him reach new customers and have affected transaction governance. Computers and the internet have helped this business escape isolation and to expand and diversify. Where orders were previously placed by phone, they are now sent via e-mail. Search engines have helped transform the business with new ideas.

This case demonstrates a positive contribution of ICTs, which had a direct impact on the producer's customer base growth through creative application: generating a website to attract new customers, participating in online courses, and using social media to communicate with producers in other parts of the world. This is a case where social media and the internet have delivered measurable disintermediation in distribution channels and encouraged an entrepreneurial escape from tradition. This case is also an example of mixed-mode governance that takes the form of longer-term cooperative agreements with a few suppliers.

8.6 Discussion and Conclusion

This study has applied concepts of transaction cost economics and social embeddedness to farming management to examine ICT usage in agricultural distribution channels. Evidence shows that ICTs can change patterns of governance. In both case studies, there has been a move to the 'middle' – banana from market, and rice from hierarchy – and in both cases a push towards longer-term contractual relationships. The effects of ICTs on distribution channels are aligned to the most efficient form of economic organization aiming to reduce asset specificity and uncertainty, and to enhance reputation and trust through an increased frequency of transactions.

There are three mechanisms of ICTs effect. Firstly, strategic, by providing access to new distribution channels through arms-length ties and disintermediation. For example, Organicos del Tropico has successfully combined arms-length's relationships with traditional distribution channels. With an entrepreneurial and business orientation, this farm has overcome isolation and increased its customer base through social media and the internet. Second, relational aims that encourage longer-term relationships. Organicos del Tropico uses Facebook as a marketplace, which facilitates access to new customers. Because of limited professional management, exploitation of social media for strategic business growth resides primarily with the producer's ability to evaluate and utilize knowledge for commercial ends. YouTube is used by the President of Banana Association to communicate his farm production capabilities to existing and new customers. Third, project management permits better control at the operational level. Both case studies illustrate the importance of negotiation and management of stock, purchase orders, relationships, and trust to successfully deliver and renew longer-term contracts. Mobile phones, e-mail, instant messaging and tracking software systems are used to communicate with customers and suppliers. These findings largely reinforce the view of hybrid modes of coordination and a move to the middle.

In rural Mexico, the main driver of rising demand for internet access has been social media, not for business, but rather for individuals to better connect with their networks. In the agricultural sector, ICTs risk remaining subservient to more traditional community principles of family, kinship and social connectivity. The pace of change is more modest in agriculture. Farmers are unable to fully exploit ICTs and find new distribution channels because they are embedded in existing business relationships. In the banana cluster, the business is more personal, with direct contact, face-to-face interactions, building reputation and trust: relational

attributes reduce transaction costs. However, in the family single estate, arm's length relationships have been more successful. Building business partnerships is important and this requires family support. Evidence demonstrates that whether sectors tend towards market or hierarchy governance, ICTs will encourage a shift to the middle emphasizing longer-term relationships and cooperation.

Facebook and other social media are new routes of communication. It might be difficult to see rapid change now, but as social media and the internet become part of daily routine, the information revolution will reach the rural sector too, also stimulated through generational change. The research detailed in this chapter demonstrates how ICTs can induce a process of change in economic organization. While their impact has been more modest that predicted, even for near-ubiquitous technology like the mobile phone, the agricultural sector in Mexico may already be on a trajectory of significant technological change. Change in human behaviour, and therefore development, takes longer, especially if it requires users to re-think the way they conduct day-to-day activities or if it alters existing modes of operation.

Notes

[1] Unión Agrícola Regional de la Sierra del Estado de Tabasco Productores de Plátano.
[2] www.elrefugio.mx (accessed 1 March 2018).
[3] Aires del Campo-Herdez sells to supermarkets, http://www.kian.com.mx/organicos.php (accessed 1 March 2018) sells to independent retailers.

References

Bakos, Y. (1998) The emerging role of electronic marketplaces on the Internet. *Communications of the ACM* 41, 35–42.
Bao, I., Huang, Y., Ma, Z., Zhang, J. and Lv, Q. (2012) On the supply chain management supported by e-commerce service platform for agreement based circulation of fruits and vegetables. *Physics Procedia* 33, 1957–1963.
Baumüller, H. (2012) Facilitating agricultural technology adoption among the poor: the role of service delivery through mobile phones. *Working Paper Series 93*, Centre for Development Research, University of Bonn, Germany.

Brush, G.J. and McIntosh, D. (2010) Factors influencing e-marketplace adoption in agricultural micro-enterprises. *International Journal of Electronic Business* 8, 405–432.

Burrell, J. and Oreglia, E. (2015) The myth of market price information: mobile phones and the application of economic knowledge in ICTD. *Economy and Society* 44, 271–292.

Cateora, P.R., Graham, J. and Gilly, M.C. (2011) *International Marketing.* Tata McGraw-Hill Education, New Delhi, India.

Chircu, A.M. and Kauffman, R.J. (1999) Strategies for internet middlemen in the intermediation/disintermediation/reintermediation cycle. *Electronic Markets* 9, 109–117.

Cloete, E. and Doens, M. (2008) B2B e-marketplace adoption in South African agriculture. *Information Technology for Development* 14, 184–196.

Coase, R.H. (1937) The nature of the firm. *Economica* 4, 386–405.

Collin, S.O. and Larsson, R. (1993) Beyond markets and hierarchies: a Swedish quest for a tripolar institutional framework. *International Studies of Management and Organization* 23, 3–12.

Cordella, A. (2009) Transaction costs and information systems: does it add up? In: Avgerou, C., Lanzara, G.F. and Willcocks, L.P. (eds) *Bricolage, Care and Information.* Palgrave Macmillan, Basingstoke, UK, pp. 386–400.

de Silva, H. and Ratnadiwakara, D. (2008) *Using ICT to reduce transaction costs in agriculture through better communication: a case-study from Sri Lanka.* LIRNEasia, Colombo, Sri Lanka, 20pp.

Granovetter, M. (1985) Economic action and social structure: the problem of embeddedness. *American Journal of Sociology*, 481–510.

Granovetter, M. and Swedberg, R. (2011) *The Sociology of Economic Life.* Westview Press, Boulder, Colorado, USA.

Grover, V., Teng, J.T. and Fiedler, K.D. (2002) Investigating the role of information technology in building buyer–supplier relationships. *Journal of the Association for Information Systems* 3, 7.

Hammersley, M. and Atkinson, P. (2007) *Ethnography: Principles in Practice.* Routledge, Abingdon, UK

INEGI (2007) Censo Agrícola, Ganadero Y Forestal. Available at: http://www.inegi.org.mx/est/contenidos/proyectos/agro/ca2007/resultados_agricola/default.aspx (accessed 29 September 2017)

Kambil, A., Nunes, P.F. and Wilson, D.(1999) Transforming the market space with all-in-one markets. *International Journal of Electronic Commerce* 3, 11–28.

Keys, E. (2005) Exploring market based development: market intermediaries and farmers in Calakmul, Mexico. *Geographical Review* 95, 24–46.

Koch, H., Leidner, D.E. and Gonzalez, E.S. (2013) Digitally enabling social networks: resolving IT–culture conflict. *Information Systems Journal* 23, 501–523.

Koch, H. and Schultze, U. (2011) Stuck in the conflicted middle: a role-theoretic perspective on B2B e-marketplaces. *MIS Quarterly* 35, 123–146.

Malone, T.W., Yates, J. and Benjamin, R.I. (1987) Electronic markets and electronic hierarchies. *Communications of the ACM* 30, 484–497.

Molony, T. (2006) ' I don't trust the phone; it always lies': trust and information and communication technologies in Tanzanian micro-and small enterprises. *Information Technologies and International Development* 3, 67–83.

Molony, T. (2009) Carving a niche: ICT, social capital, and trust in the shift from personal to impersonal trading in Tanzania. *Information Technology for Development* 15, 283–301.

Sarkar, M.B., Butler, B. and Steinfield, C. (1998) Cybermediaries in electronic marketspace: toward theory building. *Journal of Business Research* 41, 215–221.

Schultze, U. and Orlikowski, W.J. (2004) A practice perspective on technology-mediated network relations: the use of Internet-based self-serve technologies. *Information Systems Research* 15, 87–106.

Schwentesius, R. and Gómez, M.A. (2002) Supermarkets in Mexico: impacts on horticulture systems. *Development Policy Review* 20, 487–502.

Seitz, C. (2013) The impact of web-based e-commerce on channel strategy in the agricultural sector. GRIN Verlag, Munich, Germany, 14pp.

SIAP (2015) *Mexico: Food and Agricultural Atlas.* Available at: http://nube.siap.gob.mx/gobmx_publicaciones_siap/pag/2015/Agricultural-Atlas-2015 (accessed 27 February 2016).

Van Dijck, J. (2013) *The Culture of Connectivity: a Critical History of Social Media.* Oxford University Press, Oxford, UK.

Walsham, G. (1995) Interpretive case studies in IS research: nature and method. *European Journal of Information Systems* 4, 74–81.

Walsham, G. (2010) ICTs for the broader development of India: an analysis of the literature. *Electronic Journal of Information Systems in Developing Countries* 41(4), 1–20.

Wiggins, S. (2014) Rural non-farm economy: current understandings, policy options, and future possibilities. In: Hazell, P. B. and Rahman, A. (eds) *New Directions for Smallholder Agriculture*. Oxford University Press, Oxford, UK.

Williamson, O.E. (1985) *The Economic Institutions of Capitalism: Firms, Markets, Relational Contracting.* Free Press, New York, USA.

Williamson, O.E. (1991) Comparative economic organization: the analysis of discrete structural alternatives. *Administrative Science Quarterly* 269–296.

Yanyan, W. (2015) Construction of agricultural e-commerce platforms in China. *International Journal of u-and e-Service, Science and Technology* 8, 1–10.

Appendix: case study interviewees

Appendix: case study interviewees

Case ID	Respondent profile	Farming experience	Respondent education	Farm type and size
A1	Male, grower, 50+	40 years	University degree in agriculture engineering	Banana, 120 ha, 3 employees
A2	Male, grower, 40+	40 years	University degree in accounting and finance	Banana, 130 ha, cattle 500 ha, 2 employees, 130 workers
A3	Male, grower, 50+	30 years	University degree in veterinary science	Banana, 180 ha, cattle, 370 employees, 150 workers
A4	Male, grower and broker, 40+	10 years	Secondary school	Banana, 600 ha, 15 employees, 420 workers
A5	Male, grower, 19	3 years	Undergraduate in economics	Banana, 600 ha, 15 employees, 420 workers
A6	Male, grower, 29	1 year	University degree in management	Banana, 120 ha, 3 employees
A7	Male, grower, 40+	10 years	University degree in law, notary	Banana, 120 ha, 3 employees
B1	Male, grower, 30+	17 years	University degree in agricultural engineering	Organic rice and buffalo, 400 ha
B2	Male, grower, 60+	40 years	University degree in agricultural engineering	Rice, oil palm, 1,500 ha
B3	Male, internet service provider, 50+	N/A	University degree in information systems engineering	N/A
B4	Male, grower, 30+	5 years	Undergraduate degree in industrial engineering	Yuca and tuberculous plantation, 600 ha

9 Towards Alternate Theories of Change for M4ARD

Linus Kendall* and Andrew Dearden
Sheffield Hallam University, UK

9.1 Introduction

In this chapter, we explore alternative theories of change for mobile technologies for agriculture and rural development (M4ARD). Increasing attention is being given to 'theory of change' within the field of international development where it has been used by both international donors and civil society. In effect, since all interventions undertaken in M4ARD seek to generate positive change in the lives of rural populations, they can be understood as expressing some form of theory of change. While the theory of change is often only implicitly acknowledged, it is nevertheless possible to infer a theory (or theories) that link project actions to expected outcomes and impacts. Explicitly acknowledging a theory of change helps to illuminate the assumptions, conditions and processes by which an intervention seeks change. This is valuable in all phases of a project, from design to evaluation. A theory of change consists of the answer to the questions of what the context we seek to impact is, what impact is sought, how that impact will be achieved and why the proposed actions will improve the lives of the intended beneficiaries.

The way in which the theory of change is conceptualized often depends on the disciplinary outlook of the person or people assessing or evaluating the project. Unsurprisingly, authors from computer science or human–computer interaction tend to adopt technocentric approaches emphasizing technology adoption and usability, agricultural experts may focus on theories involving agricultural knowledge, while economists may adopt theories based on econometric measures. However, regardless of the framing, two popular theories of change can be observed in much M4ARD. Each of these can be interpreted as focusing on a particular understanding of 'the problem(s)' facing farmers, and so focuses on a different intervention approach. The first one, which we will call the 'market efficiency' theory, suggests that the key problem is the ability of farmers to gain a fair price for their products. In this view, adoption of mobile phones for agriculture can enable farmers to more effectively participate in markets and therefore earn a greater income for their produce.

A slightly different approach can be seen in the 'knowledge dissemination' theory, which rather focuses on the access of knowledge and training available to farmers. In these interventions – which often work directly with extension services – the goal is to provide easier access to advice, training or education via for example e-learning or telephone advice. In both cases, it is suggested that by supporting increased production and better income for farmers, ICT interventions will contribute to socioeconomic development. While we can find examples of services emphasising other issues – such as access

---

* Corresponding author e-mail: b5035879@hera.shu.ac.uk

to financial services – many, if not most, M4ARD services adopt one or both of these theories of change.

In this chapter, we will look at these two theories of change critically and highlight some of the challenges which they have faced in practice. In response to these challenges we suggest and elaborate on an alternate approach to developing a theory of change for M4ARD which draws upon human development theories. We use a case study of an ongoing research project to exemplify such an alternate theory of change. Finally, we conclude by suggesting what implications an alternate theory of change might have for the way in which we approach the design of ICTs for agriculture.

9.2 The 'Market Efficiency' Theory of Change

In a now classic study in the field of ICT4D, Jensen (2007) showed how mobile phones could lead to increased incomes for fishermen in Kerala by providing market price information and thereby reducing price dispersion[1] between different markets. He concludes that these increased incomes not only resulted in improved livelihoods but also lead to socioeconomic development through improvements in health and nutrition. 'Unlike most development projects' – he argues – mobile-phone use by farmers in his case was both self-sustaining and profitable. Jensen's study is the canonical example of the 'market efficiency' theory of change whereby the food producer, through better participation in the market, can achieve improved development outcomes.

In the decade following Jensen's study, a wide variety of M4ARD services have been employed which have sought to disseminate market information in various ways – from SMS-based platforms to interactive voice applications. Whether these interventions have yielded the aspired development impact has been the topic of several studies, revealing mixed results. In a study on an SMS-based system in India, researchers found no effect on price received by farmers, crop losses or likelihood of adopting new varieties or practices (Fafchamps and Minten, 2012). In another evaluation farmers'

perception of prices changed but this did not affect the actual price received (Camacho and Conover, 2011). Yet another study found that the interventions reviewed neither increased the number of markets farmers went to, nor did it provide farmers with higher prices (Aker and Ksoll, 2016).

At question is not just the efficacy of these interventions in achieving their sought impact but also the logic of and dynamics underpinning this theory of change. In going back to the seminal work of Jensen, Srinivasan and Burrell (2015) found that, while mobile phones certainly did provide fishermen with useful information about market prices, the process of turning price information into improved livelihoods was more complex than suggested by the initial study. First, they found that there was considerable variation between different categories of fishermen and traders in whether they could benefit from price information or not. Second, the way mobile-phone access turned to economic benefit was strongly linked with existing financial and social relationships such as those between fishermen and creditors or market intermediaries. Finally, improving income was but one of many welfare benefits that the mobile phone provided to the fishermen. Their study challenges the notion that the main challenge the mobile phone addresses in these contexts is reducing the cost of accessing information[2]. They further challenge the idea that such a process can operate without regard for socioeconomic position.

In ethnographic studies from Uganda and China (Burrell and Oreglia, 2015) there is further evidence that this theory of change fits poorly with the way farmers approach sales of their products. Similarly to the situation in Kerala, these studies show how interventions built on this theory of change face difficulties in reaching the poorest or most marginal farmers. These findings present a challenge to 'The Myth of Market Price Information' ((Burrell and Oreglia, 2015) along with the assumptions underpinning it – i.e. that farmers have limited access to information for decision making, that market price is a critical piece of decision-making information and that having access to this information will lead to better markets.

A further explanation for why this theory may be insufficient, suggested by ethnographic accounts, is that for farmers phones are primarily a 'social item' rather than an 'information delivery platform' and thus they will not readily adopt the kind of behaviour expected by this theory of change (Wyche and Steinfield, 2016).

9.3 The 'Knowledge Dissemination' Theory of Change

Aker (2011) discusses how ICTs can not only contribute to providing access to information (such as market prices and weather forecasts) but also support farmers' knowledge needs throughout the agricultural cycle. This can be through, for example, increasing access to and accountability of existing extension services or introducing new advisory, training or education services. The premise is that farmers face obstacles due to a lack of knowledge about appropriate pesticide use, fertilizer application, crop varieties and production practices. By using ICTs farmers can educate themselves, get advice from experts and other farmers and find ways to solve problems they face.

Designs of interventions for knowledge dissemination have employed a wide variety of media and modalities, such as SMS, call centres (Das *et al.*, 2012), interactive voice response (IVR) (Patel *et al.*, 2010), multimedia (Heffernan, 2007; Dearden *et al.*, 2011) and video (Gandhi *et al.*, 2007). These have largely demonstrated that such interventions are usable for even low-literate farmers.

However, evaluations of outcomes again paint a mixed picture. While impact on farmer crop choice and pesticide use has been seen, it was also shown that benefits primarily accrued to better educated and wealthier farmers (Cole and Fernando, 2012). Farmers with lower education have been found to be unable to turn knowledge or information provided into improved practice (Ali and Kumar, 2011). There is further evidence that there is a mismatch between the ways M4ARD interventions are designed to deliver advice and knowledge, and the way farmers learn and communicate about farming (Oreglia, 2013).

Aker *et al.* (2016) suggests that in the cases where ICT interventions demonstrate success this is primarily because they address a pre-existing information need and that information was a key constraint. They also make the case that demonstrated increases in farmer knowledge (such as those found by Cole and Fernando (2012) cannot automatically or reliably be turned into better yield, income or profit. Rather, they suggest, this ability is dependent upon social relationships, power relations, local market structures and other infrastructures such as roads. In other words, farmers, while appreciative of the information and new knowledge provided, depend on a wide variety of other conditions to turn that into any real economic impacts.

9.4 A 'Valued Beings and Doings' Theory of Change

Considering the challenges raised towards these theories of change, we have sought to identify an alternate theory of change and explore its implications for practice. A starting point is to move back to the broader field of development within which M4ARD sits. In the past decades, there have increasingly been moves towards development theories that expand beyond purely monetary indicators and in some cases entirely questioning the goal of economic development.

One such contemporary development theory which have been gaining prominence is the Capability Approach (CA) (Sen, 2001). Here, we suggest that the Capability Approach is an appropriate theoretical starting point for developing an alternative theory of change for M4ARD interventions. The Capability Approach has often been used to inform impact evaluation and development measurements, however in this chapter we will consider primarily how to use it generatively to formulate a workable theory of change for a M4ARD project.

In contrast with other approaches to development it is concerned with a much broader set of measures than those related to access to resources (such as income) or maximisation of welfare measures (such as happiness or fulfilment of desires). The CA is famously not a 'complete theory of change' (Robeyns, 2006) but it does provide us with a language and framework to evaluate and conceptualize the nature of the benefits we are seeking to provide, as well as a standpoint as to 'why' providing such benefits can be considered an improvement in people's lives. It is beyond the scope of this chapter to give a complete account of the Capability Approach[3], however we will seek to highlight its core concepts applicable to formulating a theory of change for M4ARD.

Central to the CA are *capabilities* – what people are *able* to do or be (Robeyns, 2005). The CA argues that development as a process should be viewed as a way to increase the *freedoms* individuals have *to achieve ways of doing or being that they have reason to value* (Sen, 2001). The CA emphasizes that freedom or *choice* is the key end goal of development. Choice is what is used to move from a set of capabilities to *functionings*, the set of freedoms that an individual chooses to realize. While capabilities can be difficult to determine, functionings are easier to observe. Operationalisations of the capability approach therefore often combine a concern of both capability enhancement as well as increased functionings (Robeyns, 2006).

As suggested above, the CA makes a clear distinction between means and ends of development: increased income, for example, is clearly a means of development. However, income is often not an end itself, rather the aspired ends towards which it contributes is capabilities – such as the being in good health – and it is these that have intrinsic value (Robeyns, 2005). An individual with a sufficient income has the capability to be in good health by choosing to use his or her income to invest in health care, thereby realizing the capability into a functioning. Means enable capabilities and functionings through a set of personal (sex, skills, education, physical condition), social (norms, practices, social hierarchies) and environmental (infrastructure, geography, climate, natural resources) conversion factors (Robeyns, 2005). Conversion factors govern the way by which a resource can be leveraged to enhance capabilities.

The capability approach places a specific emphasis on diversity through first of all its recognition of individual conversion factors and second the idea of individual choice when it comes to which capabilities to realize into functionings (Zheng, 2009). While Sen has primarily viewed capability enhancement in relation to the individual, others have made the argument for collective capabilities irreducible to the individual. Accordingly, Deneulin (2008) suggest adopting the concept of 'structures of living together'. Following this perspective, such structures are integral features of communities which cannot be understood from combinations of individual properties and over which the individual has little control. These 'structures of living together' determine both the capabilities available to the community as a whole, as well as those of individual members of the community. Considering that many M4ARD interventions operate at and seek impact on the community level – whether it is villages, farmer groups or NGOs – we suggest 'collective structures of living together' as a useful scale from which to approach capabilities.

From this brief overview of the CA we return to how it might be applied in the context of a theory of change for M4ARD. As suggested in the beginning of this chapter, a theory of change requires an answer to the questions of what impact, how we might achieve that impact and why this should be considered an improvement in the lives of those affected.

The capability approach makes a strong argument about 'why' expansion of substantive freedoms improves the lives of those involved. However, the exact choice of these freedoms, i.e. 'what impact', should be up to a process of social negotiation in the target communities. This implies that rather than prescribing specific forms of livelihood improvement; we should be seeking to gather participants' views of the desirability of various outcomes. Therefore, applying the CA creates a demand for an in-depth exploration of the values of participants in the project. On the scale at which many M4ARD applications operate on, approaches grounded in participatory development may be suitable ways of achieving this (Alkire, 2005).

When it comes to the question of 'how', we can view ICTs as one of many resources or means available to an individual or community that can be used towards capability achievement. A focus on capabilities and functionings does not suggest that resources – such as ICTs or income – are unimportant, but rather they should be considered as 'instruments' and enablers (Robeyns, 2005; Zheng, 2009). Resources such as ICTs are drawn upon to shape and enable both individual capabilities as well as 'structures of living together'. They can directly result in capability enhancement, as in the case of mobile phones enabling participation in community life by providing long-distance connection. They can also interact with other resources that can then in turn be used for capability enhancement – for example by enabling greater income through better participation in the market. To understand how this process happens, however, we need to pay special attention to conversion factors – both personal, social and environmental. When it comes to ICTs this moves beyond accessibility and affordability to also include, for example, skills to use the ICT or understanding of the opportunities ICTs provide (Kleine 2010).

In summary, a theory of change informed by the Capability Approach would have the following hallmarks:

- The ultimate impact of a development intervention should be the expansion of freedoms to achieve 'beings and doings that people have reason to value'.
- This expansion is most commonly observed through the proxy of functionings – the capabilities that individuals choose to realize.
- What these valued beings and doings are, is to be defined through a participatory process involving those affected by the change.
- The impact on capabilities for individuals and communities of ICTs are mediated by existing individual conversion factors as well as structure of living together.
- New communication technologies influence and result in new structures of living together and may further contribute to capabilities as conversion factors.

This suggests that to apply this as an approach for M4ARD intervention, it requires:

- A participatory exploration and elaboration of the values of the communities we work with.
- Concern with, and incorporation of, pre-existing conversion factors and structures of living together that shape the evolution and impact of introduced technologies.
- A recognition of the complexity and uncertainty inherent in an approach that considers diverse local structures of living together, conversion factors and individual agency.

This leads to a distinctly different theory of change from the two introduced in the beginning of the chapter. It is a theory that does not negate the importance of market information, market efficiencies or knowledge dissemination. However, it incorporates these into a broader understanding of the ways in which these are applied, as well as what development they can be considered as contributing to. To illustrate this in practice, we will be using an ongoing research project on agricultural development in West Bengal, India where our starting point for the introduction of an M4ARD intervention has been the identification of 'valued beings and doings'.

9.5 'Valued Beings and Doings' in Practice

We will use work begun in 2016 with a non-profit in West Bengal in India to exemplify our proposal. The non-profit we are working with has for 30 years supported marginal and smallholder farmers in the region to adopt agricultural practices that are ecologically, economically and socially sustainable.

Considering the Capability Approach had an important influence on the starting point for our research project, instead of asking 'what are the issues the organization and its stakeholders face?' we sought to understand 'who are those involved in the organization and what are the ways of doing and being that they value?'. Drawing on the idea of 'structures of living together', special emphasis is placed on structures of communication and interaction within the organization and between the organization, groups of farmers and other stakeholders.

To incorporate this study into a project that is also concerned with the design and implementation of appropriate information technologies, we adopted action research as our approach. As an initial step, an action learning set with members from different parts of the organization was engaged. Their role was to help direct the inquiry as well as discuss and analyse outputs from the research. From this followed a six-month period of ethnographic observation, interviews and group discussions primarily centred on two field offices as well as the head office of the organization.

9.5.1 Structures of living together

From the research project conducted with the organization, several structures of living together were identified. The process by which these were identified consisted of thematic analysis of the raw data followed by reflection and critical discussion of these analyses with the action learning set. That these are indeed valued beings and doings was triangulated through: multiple sources in various parts of the organization including farmers, field workers and head office staff; as well as through multiple methods, including interviews, observation and group discussion.

9.5.2 Being resilient and self-sufficiency

The ability of the agricultural system to be resilient was an important value expressed. This included both the resilience of the agricultural communities as well as of the organizational structures designed to support them. This value was expressed both in the content of the organization's activities as well as in the way they structured their work.

When it comes to content, they adopted an approach to agriculture founded on agro-ecological principles (Altieri and Nicholls, 2005). Agro-ecology places the long-term sustainability of the farming system at the centre, suggesting that agricultural practice has increasingly chosen to focus on short-term yields, achieved through practices which are harmful for the long-term ability of the farming system to guarantee farmer livelihoods. As an example of this, the organization had worked with the farmers to identify alternative summer crops to hybrid rice varieties which did not require irrigation, fertilizers or pesticides. By adopting summer crops such as lentils, the farmers could not only avoid investment in labour and inputs, but would also see improved nutrition. Investments in programmes such as local-variety seed saving, farm-level water-harvesting structures and kitchen gardens were all designed to allow the village, the farm and the family respectively, to have greater resilience and self-sufficiency.

In terms of their organizational structure this was reflected in the fact that each local office

was intended to be a separate entity from the head office with its own governance structure and funds. While this independence greatly varied from field office to field office, one of the ways that self-sufficiency and resilience was maintained was through the way in which financial support was provided. While the head office raised the greater part of the funds, for individual investments, these were provided to a fund held by the local farmer's group. That fund in turn, rather than providing funds as a 'grant' to the beneficiary, provided it in the form of an 'investment' which the beneficiary eventually should pay back to the farmer group fund. In this way, the farmer groups could eventually hold a revolving fund that could be used to continuously invest in improvement projects on individual farms.

9.5.3 Engaging holistically and long-term

Valued ways of being such as resilience and self-sufficiency are interrelated with a desire and need for holistic and long-term engagement. For a farm to become sustainable, many years of engagement is required. It is also not sufficient to engage with just a single farm, as neighbouring farms will influence it through both ecological as well as social interactions. The organization had from the start engaged with communities in this way, focusing on deep connections with a few communities rather than shallow links across many. Many of the senior members of staff of the organization had spent extensive amounts of time living and working in their project areas.

Holistic engagement meant addressing different topical concerns – such as including food habits as part of their intervention – but also operating on different interlocking scales: the individual farmer, the village and the block[4]. Drawing on the previous example, adoption of alternate summer crops such as lentils instead of rice involved engaging with and challenging strongly held norms. That rice has a special place in the local culture is exemplified by the phrase 'rice sleep' – the desired result of, and a state of well-being following, a meal with a large quantity of rice. Growing an additional rice crop, even if less beneficial financially or from a perspective of sustainability, might be valued in part because rice (specifically) provided a particular sense of food security.

When it comes to technologies, an example can be drawn from a trial of an IVR-based questions and answers system we piloted with the organization. In the evaluation of the project, one of the main trainers said that while he felt that it was a valuable tool he suggested that questions and answers as a means of conveying information was inappropriate. Farmers would use the service to ask reactive questions based on pest or disease problems expecting an answer in the form of a 'medicine' to apply. However, the trainer suggested – based on agro-ecological principles – that the pest was merely a symptom of a problem that could be solved only by long-term redesign of the farm. He felt that such holistic considerations were difficult to combine with the format of a Q&A system.

9.5.4 Close social interactions

That social interactions were important in the work of the organization and its engagement with the farmers became most evident in the activities of the field office.

For example, a group of experienced farmers were observed being trained on a topic that all of them were well familiarized with. They were neither given financial inducements nor other direct material benefits from attending. When later asked, the trainer explained that the real reason for hosting these meetings was to create a social meeting space where not only agriculture but also a wide variety of other topics could be discussed. He suggested that villages previously had these meeting spaces, but today they were increasingly rare and one of the contributions of the organization was to reinstate them.

The work of many of the organization's field officers freely mixed social and work oriented interactions with relatively little distinction between different types of relationships. Looking at how this affected technology use, one of the main technologies observed in use was WhatsApp. Several uses of the messenger application appeared inefficient and more amenable to structured workflow tools. However, when

seen through the lens of their work practices, an unstructured, social messaging application seems ideally suited.

9.5.5 Reasons to value?

These three structures of living together were all identified through the work of the organization by staff members and farmers we interacted with during this part of the research project. That they are valued ways of doing or being which the participants in the context see as relevant makes them important to consider for an intervention.

However, these valued beings and doings are by no means the only ones the organization and its stakeholders aspire to, and they were subject to active disagreement throughout the research. The process of choosing the above three values as foundational for our theory of change rests on the Capability Approach argument that, for a freedom to be 'substantial', it must be something that the individual or community has 'reason to value'. In the case of the values mentioned above we can identify clear reasons why these are valued.

To give some examples, the resilience of the social networks in the context allows for communities which better can respond to external challenges such as climate change. As such, social capital is an important asset to draw upon in crises and can be directly linked to agriculture through practices such as community sharing of crop genetic diversity. Long-term approaches recognize that sustainable agricultural approaches that are embedded in local conditions take time to develop. The way in which agricultural practice is interlinked with all aspects of rural life in this region speaks for holistic engagement.

We recognize that choosing to incorporate these as the foundation for our theory of change is an act of choosing some and suggesting that there is 'greater' reason to value these rather than others. However, we do not view this as necessarily a weakness of the approach presented in this chapter, rather we see a strength in the explicit acknowledgement that adopting a theory of change is a normative exercise. Our justification lies in the participatory processes and open dialogue within the action learning set and with the organization in developing the project approach.

9.6 Identifying the Role of ICTs

In developing our alternate theory of change, we suggest that an inquiry into structures of living together – such as the one described above – forms the basis for the intervention. These 'valued beings and doings' thus become the core of 'what' our intervention should seek to support or enable. The next step is to identify the ways (if any) that ICTs might contribute effectively to them. At the time of writing the project and the action learning set are planning possible intervention strategies. Below, we suggest some of the ways that the structures of living together discussed above might be incorporated into an M4ARD strategy.

9.6.1 Being resilient and self-sufficient

The goal of resilience and self-sufficiency suggests a role for ICTs which builds on assets which the community already has access to or can gain access to by themselves. As an example of this, we can take a case of a project, studied as part of this research programme, to share weather data to villagers and field workers from reports provided weekly in one of the field offices. While the project plan was to disseminate the data by SMS – for which a large number of mobile-phone numbers were collected – in practice this did not work because the organization lacked the capability to manage the online SMS sending system. Furthermore, recipients had phones without the functionality to display Bengali language SMS. Instead, the dissemination of weather reports via phone consisted of an informally created WhatsApp group where one officer shared photographs of the weather reports with others, who subsequently transcribed the reports onto community notice boards. This suggests that one approach would be to seek to develop new ways of employing technologies that the organization already has access to, rather than introducing new ones.

Another way of viewing this can be to look at it as ways of building on the conversion factors which either limit or enable conversion of already existing technologies into capabilities, or limit or enable appropriation of new technologies. Working with participants in the communities as technology stewards or ethnographic action researchers are ways in which we can seek to build that local technological self-sufficiency and resilience.

9.6.2 Engaging holistically and long-term

One of the obstacles that was often raised in discussions about the values above was the fact that they were increasingly dependent on a growing number of external funding agencies. These funding agencies brought with them not only shorter-term timelines (3–5 years) but also specific, outcome-oriented project goals which in many occasions stood in conflict with the type of long-term, holistic engagements that the organization valued. The greater number of projects led to additional recruitment of staff members who did not necessarily share the intention or aspiration for a long-term, deeply embedded engagement with the target communities.

Certain values and ways of being are incorporated into the structure and design of many ICTs that can be detrimental to this value. One example is the commonly adopted modality of 'questions and answers' which one staff member suggested as unsuitable for communicating the more holistic ways in which sustainable agriculture would approach solving problems such as those of pests and plant disease. As an alternative, technologies that aid in or support the maintenance of community meetings (rather than seek to supplant them) may help contribute to long-term and holistic ongoing dialogue involving all members of the community as a group. This stands in contrast to more individualized approaches viewing the individual farmer as a user of the system.

Increasing emphasis on project work, reporting and measurement was highlighted by another staff member as having resulted in more specialized and disconnected project teams. These in turn led to less holistic engagement of staff in the issues the organization cared about. For example, previously held film screenings on topics relevant to their work had disappeared. He also noted how many of the things they considered important to their work, for example the emergence of empowered lead farmers in the villages, fit poorly with measuring systems introduced by funders oriented towards outcomes and quantitative impact assessment. Therefore, we might seek to adopt modalities – such as sharing videos or organizing film screenings – which address sharing of values or allow certain individuals to take up a leadership role in front of others in the community.

For development practitioners, such as those in this case, changes to the way impact is assessed can mean a move away from valued holistic ways of being towards a practice that emphasizes reduction, calculation and measurement. ICTs are often well aligned with such practices and therefore special care must be taken in cases like these not to further marginalize such values (Hayes and Westrup, 2012; Hayes, 2015). Conscious effort is needed to identify alternate ways in which ICTs can operate in these communities.

9.6.3 Close social interactions

We can recognize that M4ARD interventions such as Digital Green (Gandhi *et al.*, 2007), Kheti (Dearden *et al.*, 2011) and Avaaj Otalo (Patel *et al.*, 2010) all highlight the importance of the social aspects of phone use in their interventions. Aforementioned ethnographic studies (Oreglia, 2013; Wyche and Steinfield, 2016) further suggest this applies more broadly in the process of agricultural knowledge sharing and ICT use among farmers. Even if the most direct benefits of access to a mobile phone are economic, several studies conclude that the most *valued* impact is the ability 'to communicate and be connected' (Smith *et al.*, 2011).

This corresponds well to the values expressed by the staff members we worked with in this case. In their use of technologies, we saw preference for communicative and social technologies such as WhatsApp above those that might be more efficient for information sharing. A theory of change in this project needs to be built on the basis of, and ideally strengthen, the social and communicative interactions between the participants. It calls for an emphasis away from a direct focus on delivering and disseminating specific discrete pieces of knowledge and information towards building tools that allow the community to better communicate and build social relationships. Additionally, these discussions suggest that technology adoption and change in this context is, and by necessity needs, to be closely linked with organizational and social change.

9.6.4 A theory of change for our project

With these reflections in hand, we can return to the definition in the beginning of this chapter of a theory of change as a framing of the situation, a definition of what impact we will achieve and why this will improve the lives of people we are working with.

From our findings, we can see that the desired capabilities relate to resilience, holistic and long-term engagement and work conducted through close social interactions. This does not suggest that sharing price information or disseminating knowledge should not be part of our intervention. However, such interventions should be built in ways that incorporate a 'structure of living together' that supports, rather than detracts, from desired capabilities. This could mean, for example, technologies that enable face-to-face meetings as opposed to reducing the need for them. It could mean developing impact assessment tools that allow for sharing of more holistic impacts – for example through video. It would highlight the need to recognize and build on existing conversion factors, such as ICT skills or ability to critically evaluate technologies offered by external parties, to support self-sufficiency when it comes to technology adoption.

Supporting the organization and its stakeholders to strengthen their ability to enact these desired ways of being and doing, maintaining it in their 'structure of living together', would not only (in our theory) enable them to improve their agricultural outcomes but would also constitute a 'development end' in the conceptualisation of the capability approach. This may seem to be a rather indirect role for ICTs to play in this context. We argue that locating ICTs in such a way is a strength of this approach, reducing the risk of an overly expansive or optimistic view of the role that ICTs play in the development process (Zheng, 2009).

9.6.5 Broader implications for M4ARD strategy

While the concerns and strategies presented in the previous section almost certainly have parallels in many other contexts, they are contingent on the specific context within which we have worked. However, we argue that there is scope for drawing more general implications for M4ARD interventions. Specifically, this approach provides an outline of a method by which to construct a theory of change that moves beyond both a narrow economic view of development as well as one which places insufficient attention to the process by which knowledge turn into practice.

First, this approach places emphasis on the actors involved in the practice of development. Rather than starting from a perceived problem or 'issue' and seeking ways for ICTs to contribute to it, we look at the actors, what are their valued ways of being and doing and how we might contribute to strengthen these or remove obstacles for them. Recognizing that it is challenging to articulate such valued beings and doings, we suggest that a potential role for researchers and practitioners in this field can be in facilitating the processes and spaces that allow for their identification and expression. This can be done through utilizing ethnographic methods and skills or drawing upon the rich methodological space of participatory development.

Second, this theory of change foregrounds the 'socially embedded' role for technology (Avgerou, 2010) highlighting the relationships between social interactions and organizational structure and change. An M4ARD project viewed from this lens cannot have technology as its sole focus, but will be linked to and dependent upon other developments within the social and organizational system it is placed within. It will recognize that technological change is linked with social and organizational change, and adopt suitable approaches to handle that. One way in which we have sought to do so in this project is through an explicit combination of action research and participatory design. Engaging members from the organization as part of an action learning set allows linking the activities of the project with broader changes happening in the organization. As technology designers, we are unlikely to be the main social and organizational change agents. Recognising the deep inter-linkages between technological, social and organizational change, therefore requires approaches whereby we seek to enable those who can serve as such change agents to appropriate and apply technology in suitable ways. Methodologies such as Ethnographic Action Research, Participatory Design or previously mentioned Technology Stewardship are all ways in which this can be achieved.

9.7 Conclusion

We have discussed the theories of change adopted in M4ARD projects, highlighting the challenges facing the often implicitly accepted theories of 'market efficiency' or 'knowledge dissemination'. From this discussion, we have suggested that M4ARD projects need to both make explicit, and critically examine their underlying theory of change. As a response, we have proposed the Capability Approach as a theoretical grounding for an alternative theory of change for M4ARD. In formulating an M4ARD technology strategy, such an approach foregrounds the actors involved in the development process, reviewing their 'valued ways of being and doing'. It also suggests assessing individual 'conversion factors' and building upon collective 'structures of living together' that shape and in turn may be influenced by technology.

Adopting an approach influenced by the CA meant seeking a local theory of change that builds on values gathered through participatory engagement. An emphasis on 'valued ways of being and doing' can uncover important insights about both the goals as well as means of M4ARD projects. To illustrate this, we presented a case study where the requirements of communities adopting sustainable and agro-ecological agriculture argued for specific approaches to ICTs. These approaches emphasized support for existing collaborative and communicative practices as opposed to disseminating disjointed packets of information to isolated individuals. While information dissemination is valuable, in our case when instantiated within specific ICT systems designs, it can conflict with ways of being and doing that the community has reason to value.

Notes

[1] Price dispersion refers to the difference in price received for the same good across separate markets.

[2] The notion of reducing 'search costs' for both public (market price, agricultural knowledge) and private information (Aker 2011).

[3] For a more in-depth accounts see for example Sen (2001), Robeyns (2005, 2006), Deneulin (2006).

[4] A block is an administrative area of India that can be compared to county. A block is a subdivision of a District, and a District is a subdivision of a State. A block will generally include many villages and settlements distributed over an area of a few hundred km[2].

References

Aker, J.C. (2011) Dial 'A' for agriculture: a review of information and communication technologies for agricultural extension in developing countries. *Agricultural Economics* 42, 631–647.

Aker, J.C. and Ksoll, C. (2016) Can mobile phones improve agricultural outcomes? Evidence from a randomized experiment in Niger. *Food Policy* 60, 44–51.

Aker, J.C., Ghosh, I. and Burrell, J. (2016) The promise (and pitfalls) of ICT for agriculture initiatives. *Agricultural Economics* 47, 35–48.

Ali, J. and Kumar, S. (2011) Information and communication technologies (ICTs) and farmers' decision-making across the agricultural supply chain. *International Journal of Information Management* 31, 149–159.

Alkire, S. (2005) *Valuing Freedoms: Sen's Capability Approach and Poverty Reduction.* Oxford University Press, Oxford, UK.

Altieri, M.A. and Nicholls, C.I. (2005) *Agroecology and the Search for a Truly Sustainable Agriculture.* United Nations Environmental Programme, Mexico City, Mexico.

Avgerou, C. (2010) Discourses on ICT and development. *Information Technology and International Development* 6, 1–18.

Burrell, J. and Oreglia, E. (2015) The myth of market price information: mobile phones and the application of economic knowledge in ICTD. *Economy and Society* 44, 271–292.

Camacho, A. and Conover, E. (2011) The impact of receiving price and climate information in the agricultural sector. *IDB Working Paper Series*, 220, Inter-American Development Bank, Washington DC, USA.

Cole, S.A. and Nilesh Fernando, A. (2012) The value of advice: evidence from mobile phone-based agricultural extension. *Harvard Business School Finance Working Paper*, No. 13-047.

Das, A., Basu, D. and Goswami, R. (2012) Accessing agricultural information through mobile phones: lessons of IKSL services in West Bengal. *Indian Research Journal of Extension Education* 12, 102–7. Available at: http://seea.org.in/vol12-3-2012/19.pdf (accessed 7 April 2014).

Dearden, A., Matthews, P. and Rizvi, H. (2011) Kheti: mobile multimedia in an agricultural co-operative. *Personal Ubiquitous Computing* 15, 597–607.

Deneulin, S. (2006) *The Capability Approach and the Praxis of Development.* Palgrave Macmillan, London, UK.

Deneulin, S. (2008) Beyond individual freedom and agency: structures of living together in Sen's capability approach to development. In: Comim, F. *et al.* (eds) *Capability Approach: Concepts, Measures and Application.* Cambridge University Press, Cambridge, UK, pp. 105–124.

Fafchamps, M. and Minten, B. (2012) Impact of SMS-based agricultural information on Indian farmers. *The World Bank Economic Review* 26, 383–414.

Gandhi, R., Veeraraghavan, R., Toyama, K. and Ramprasad, V. (2007) Digital green: participatory video for agricultural extension. In: *International Conference on Information and Communication Technologies and Development, ICTD 2007, Bangalore, India,* IEEE. Piscataway, New Jersey, USA.

Hayes, N. (2015) Governing impact in the international development sector, Introduction: NGOs and Impact. *Key note, 13th International Conference on Social Implications of Computers in Developing Countries,* Negombo, Sri Lanka, 2015. Available at: http://www.ifipwg94.org.uk/files/Keynote_-_NiallHayes.pdf (accessed September 2017).

Hayes, N. and Westrup, C. (2012) Power/knowledge and impact assessment: creating new spaces for expertise in international development. *New Technology Work and Employment* 27, 9–22.

Heffernan, C. (2007) The Livestock Guru: the design and testing of a tool for knowledge transfer among the poor. *Information Technology and International Development* 4, 113–121.

Jensen, R. (2007) The digital provide: information, market performance, and welfare in the South Indian fisheries sector. *The Quarterly Journal of Economics* CXXII (2007), 879–924.

Kleine, D. (2010) ICT4What? – Using the choice framework to operationalize the capability approach to development. *Journal of International Development* 22(5), 674–692.

Oreglia, E. (2013) When technology doesn't fit: information sharing practices among farmers in rural China' In: *Proceedings of the Sixth International Conference on Information and Communication Technologies and Development, ICTD '13, Cape Town, South Africa*, ACM Press, New York, USA. pp. 165–76. Available at: https://doi.org/10.1145/2516604.2516610 (accessed 15 June 2016).

Patel, N., Chittamuru, D., Jain, A., Dave, P. and Parikh, T.S. (2010) Avaaj Otalo: a field study of an interactive voice forum for small farmers in rural India. In: *Proceedings of the 28th International Conference on Human Factors in Computing Systems, Atlanta, Georgia, USA*. ACM Press, New York, USA, pp. 733–742. Available at: https://doi.org/10.1145/1753326.1753434 (accessed 15 June 2016).

Robeyns, I. (2005) The capability approach: a theoretical survey. *Journal of Human Development* 6, 93–117.

Robeyns, I. (2006) The capability approach in practice. *Journal of Political Philosophy* 14, 351–376.

Sen, A. (2001) *Development as Freedom*. Oxford University Press, Oxford, UK

Smith, M.L., Spence, R. and Rashid, A.T. (2011) Mobile phones and expanding human capabilities. *Information Technologies and International Development* 7, 77–88.

Srinivasan, J. and Burrell, J. (2015) On the importance of price information to fishers and to economists: revisiting mobile phone use among fishers in Kerala. *Information Technologies and International Development* 11, 57–70.

Wyche, S. and Steinfield, C. (2016) Why don't farmers use cell phones to access market prices? Technology affordances and barriers to market information services adoption in rural Kenya. *Information Technology for Development* 22, 320–333.

Zheng, Y. (2009) Different spaces for e-development: what can we learn from the capability approach? *Information Technology for Development* 15(2), 66–82.

10 Mobile for Agriculture (m4Agric) Services: Evidence from East Africa

Richard Duncombe*

Centre for Development Informatics, University of Manchester, UK

10.1 Introduction

In recent years, the potential benefits of mobile phones for agricultural development have been propagated by key donors (Qiang *et al.*, 2011; World Bank, 2011; GSMA, 2015) and demonstrated through widespread application of new mobile-for-agriculture (M4Agric) innovations. Mobile phones provide coverage and connectivity, but also embody additional functionalities, acting as platforms for applications (e.g., knowledge repositories, farmer helplines and mobile finance), linking farmers via networks to the computer-based information systems of larger service providers. The GSMA (Global System Mobile Association) 'mobile4development tracker' tracks the progress of 98 live M4Agric deployments – those which are most visible, but probably only a smallish proportion of the total Agricultural Value Added Services (AgriVAS) worldwide.[1]

Africa has the largest number (52), with service providers fairly evenly split between Mobile Network Operators (MNOs), Non-government Organizations (NGOs), VAS Providers, Technology Vendors and Government. The number of future users has been forecast to reach 80 million by 2020, with 30 million in Africa. The GSMA estimates, across the countries in their analysis, that 35% of agricultural workers in developing countries are potential mobile subscribers, of which 22% are potential AgriVAS users. Key drivers are identified as accelerated rollout of mobile networks in rural areas, the launch of more AgriVAS solutions, and increased awareness of the benefits of Agri-VAS, such as supply chain efficiency and income growth, which is seen as the main reason for users' propensity to purchase services (GSMA, 2015).

This chapter will suggest that the role of mobile-phone-led services needs to be viewed in an enabling context that highlights both systemic and organizational change. The chapter will investigate this by focusing on a case study, tracking the evolution of a mobile-phone-led service from East Africa. The findings will seek to inform the enactment of such interventions in resource-poor environments, where there is a strong need to link policy to the requirements of strategically important sectors such as agriculture (Poulton *et al.*, 2010; *World Development Special Issue*, 2010).

10.2 M4Agric Services

Impact studies of From Africa are limited. Subervie (2011) evaluated the impact of SMS-based alerts for farmers via *Esoko* (the Swahili word 'Soko' meaning market), a for-profit company providing AgriVAS for smallholder farmers.

---

* E-mail: richard.duncombe@manchester.ac.uk

They used econometric modelling of spatial arbitrage conditions, and found a significant effect on yam prices with a 10% increase amongst the treatment group of 500 farmers to whom mobile phones were distributed in the northern region of Ghana. However, more recent data for *Esoko* found that . . .

> the treatment effect disappears in the second year of the study, and is never present for other crops (maize, cassava, and groundnut). The price alerts had some effects on farmers' choices about where and when to sell, particularly for yam, but also for other crops. However, the main mechanism through which the price alerts led to increases in yam prices was by improving farmers' bargaining power with traders.
>
> (Hildebrandt *et al.*, 2014, p. 1)

In a similar vein, Van Campenhout (2012) used double-difference and fixed-effects methods to evaluate the impact of the Grameen Community Knowledge Worker (CKW) Programme in the Mount Elgon region of Uganda. The programme claims to reach the most isolated rural villages by employing a network of local advisors drawn from the communities they serve. These Community Knowledge Workers – farmers themselves – use smartphone applications to give fellow farmers information on weather and market prices and advice on caring for their crops and animals and treating pests and diseases (CKW, 2016). The study looked at changes in farmer knowledge, attitudes towards information, changes in farming practices, and outcomes. Overall, knowledge and attitude towards information was improved (as you might expect) and practices were changed, such as use of crop spacing and manure. As regards outcomes, a significant increase in the price received for maize was recorded, but no 'convincing effect for productivity' (Van Campenhout, 2012, p. 21). The explanation for this was the difficulty of measuring inputs reliably, and the time period (probably measured in years) before real productivity improvements could be realized (e.g., due to crop spacing and manure not having immediate effect on soil fertility).

Given the uncertain and limited evidence of positive outcomes, a number of researchers have suggested reasons why there appears to be a growing gap between the expected benefits of mobile-phone-led services and the empirical evidence (Burrell and Oreglia, 2015; Duncombe,

2016). Two constraints stand out. First, (mis) management of information and technology which can result in both additional costs and unrealized gains. ICT involves complexity in design, maintenance and management systems. M4Agric projects, in common with most ICT4D projects, experience cost overruns, delays in implementation, rapid obsolescence of hardware and software, and other problems of compatibility and security (Heeks, 2010). Many of these constraints are magnified in the rural areas of developing countries, from where farmers wish to connect digitally. Second, the significant time-lags between investment in new technologies and productivity/output improvements. During these time lags, it is necessary to make sufficient complementary investments in other, more critical, factors of production. In the case of agriculture in sub-Saharan Africa these would include investments in infrastructure such as electricity and roads, sufficient financial investment to upgrade agricultural production technologies, and the ability to access markets large enough to warrant the increase in production. To conclude, extracting productivity benefits from ICT requires many complementary investments; it also requires changes in complementary processes and structures (i.e. just changing the technology is insufficient) and the process of change and transformation can take many years.

10.3 Field Research

The research strategy chosen was a case study approach (Yin, 2003), which is used to define the organizational boundaries for the research setting according to the involvement of *primary* and *secondary stakeholders*. The research was designed to provide contrasting stakeholder views and was *instrumental* in the sense that it provides a framework for investigating the connection between change in agricultural systems and stakeholder management practice (Stake, 2005). A categorization of stakeholders was employed as shown in Table 10.1 below. Primary stakeholders are those which are directly involved with the day-to-day management and development of the mAgric application (the service providers, agricultural intermediaries and

Table 10.1. Stakeholders in the agricultural value chain.

Secondary stakeholders	

Local organizations	International organizations
• Local NGO	• International NGO (INGO)
• Local consultant	• Donor
• Private sector organization	
• Government organization	
• Educational establishment (inc, Universities)	
• Mobile service provider	

Primary stakeholders	
• Agricultural service provider	• Agricultural information content provider
• Agricultural intermediary (traders, extension workers, farmers' groups, etc)	• Agricultural knowledge broker
• Farmers	

the farmers). The secondary stakeholders are those which have some intermittent involvement (such as consultants) or those which have an interest in the outcomes associated with the intervention (e.g. government or donors).

The data collection methods consisted of key informant interviews and field observations that were conducted during March and April 2014 (Appendix 1). Data were collected from multiple stakeholders (primary sources) as well as from documented secondary sources which included official reports and institutional websites. This enabled verification and contrasting of the data collected. In-depth interviews were able to reveal a detailed story of the respondents' experiences in applying new technologies to the agricultural sector over a number of years. A number of days were also spent observing the mAgric systems in practice, both in the back office operation, and out in the field, which involved the researcher spending two days moving between farms with agronomists, observing the practical application of the mobile-enabled systems.

The case selection was that of m-Farm based in Kenya. This case was selected because it represented an example of an mAgric intervention that had been operational for sufficient time to observe a number of transitions in technical, organizational and business parameters. m-Farm is a social enterprise that provides farming information and group-based trading opportunities favourable to small farmers, as well as subscription-based daily agriculture/horticulture pricing information in all major cities and

markets in Kenya. m-Farm was created in 2010 after attracting US$100,000 (£65,000) of seed funding from TechforTrade, a UK charity focused on bridging the divide between emerging technology, international trade and economic development.[2] m-Farm's founder – Jamila Abass, a 29 year-old computer scientist – recognized that lack of pricing transparency meant that farmers did not always get the best prices for their produce. For many small, low-volume farmers, the only source of information about the market rate for crops came from the traders who were trying to buy from them. Abass described them as:

> Oppressed for decades and disconnected in terms of information . . . many farmers only have the produce, but don't have the means to market their produce themselves . . . they have to rely on middlemen who show up and give them both the price and the buyer. They have no information and no alternative market. We wanted to close that information gap between the farmers and the market.
>
> (Solon, 2013)

10.4 Case Study: m-Farm

m-Farm sought to expand access to market information by providing up-to-date market prices via a mobile app, direct to farmers, whilst also connecting farmers with buyers, and cutting out trading middlemen. m-Farm developed a tool that allowed farmers to SMS the number 20255 to receive information relating to the retail price of their products – updated daily with information for 42 crops (including peas, sugar snaps, avocadoes, passion fruit, peanuts, potatoes, cassava and mangos) sold in five local markets. m-Farm moved to integrate transactions and payments, which were added to the system in 2012, handled by m-Farm's integrated mobile money transfer system, built on the m-Pesa platform, with the ability to connect into buyers' and farmers' bank accounts (if they had them). m-Pesa is the dominant mobile money service provider in Kenya, operated by Safaricom. Safaricom has opened up the m-Pesa platform to allow local and international developers to further integrate value-adding applications, through the provision of an Application Programming Interface (API) which makes it easier to integrate with other financial and information services.[3] Experience of the emerging system by the farmers was largely positive, but m-Farm soon realized that the technology was also having a disruptive effect, creating a problem . . .

> farmers don't have storage facilities and they know that the buyer who comes around to the farm can just go next door and get produce from someone else. So we could end up taking away the only access to market they have . . . others have come to the market using technology to create a trading platform that farmers are not ready for. They have also been set up by non-profit organizations and run out of money, leaving the farmers high and dry. This makes them sceptical.
> (Abass, quoted in Solon, 2013)

m-Farm realized that the problem faced was not primarily price transparency, but the fact that farmers were producing in low volume and that buyers in urban centres were not inclined to source the volume they need from multiple small-scale producers. This led m-Farm to conclude that a more fundamental change was required in the way that the farmers produced and marketed their produce.

This led m-Farm to develop a group-selling tool that provided a platform for collaboration, whereby small farmers were able to bring produce to drop-off points. They were then able to send an SMS to the system advertising (collectively and in larger quantities) what they have to sell. When an order was placed through m-Farm, the farmer took his or her produce to a designated collection point and sent a message to

confirm the produce had been delivered. The buyer then collected the produce and verified the quantity and quality by sending a message to m-Farm. Once that had been confirmed, and the order had been fulfilled, the money was released by m-Farm to the farmer's account. With larger orders, where multiple farmers were involved, the money was distributed between different accounts. m-Farm also addressed problems related to inputs, as seen from the following comments from participants:

> We realized that the production side was a much deeper problem than the market. Why? Because these farmers were using old ways of doing things – they needed education on new methods and ways of working the land . . . when they have pests or diseases the first line of defence is the local Agrovet – but they lacked information on why they are buying products and how they should use them?
> (university-based stakeholder, 02/04/2014)

Additionally:

> . . . farmers did not have access to money to buy the products upfront, so they only buy a minimum amount of the product (e.g., pesticide/fertilizer) and spray it on a wide area of land, and find out that this product appears not to be working . . .
> (mFarm manager, 04/04/2014)

These problems stimulated the development of a group buying tool, allowing farmers to pool resources to negotiate better prices for farm inputs. By this time, m-Farm was starting to transform into a for-profit organization, taking a transaction fee for every deal done using its platform. This allowed it to grow the number of users from 2000 in early 2012 to 7000 in 2014. A study in central Kenya with 600 farmers showed that farmers could double the volume of their sales by using m-Farm . . .

> m-Farm can lower costs [of supplies] and offer better margins for farmers, but the other value proposition is a consistent market . . . It's not just about the prices but also knowing if a buyer will be available
> (Nairobi iHub manager, 28/03/2014)

m-Farm faced further obstacles in relation to the activities of existing government-run extension officers that were providing services in the same locations as m-Farm were operating:

The problem is that the extension worker is the one moving around all the farmers, and they used to get all that information from farm input suppliers – big companies that supply the Agrovets who just want to push their own products . . . like Sygenta . . . who train farmers in using their own products.

(mFarm manager, 04/04/14)

m-Farm realized that contact with farmers via their mobile application was not sufficient, they needed their own agents to interact with farmers as well as to close deals:

We decided to come up with a new model. We employed agents that work only with the farmers . . . relationships with the farmers and the buyers then changed completely. The agents are the people who were working with the farmers directly, and they would be the eyes of m-Farm on the ground – their work was to recruit, to train and to check on the quality of the produce when it was harvested. The downfall was side-selling. They would feel more powerful than the farmers and cut them off and determine whether they sell or not, so we realized that the more money they made out of commissions, the more powerful they became.

(mFarm manager, 04/04/14)

This prompted another change of tack and after further consultations with the farmers, they realized they needed to deal with trusted agents who were not primarily business-orientated:

The farmers wanted to work with people who could provide more agronomical information – they would put more trust in them. We changed the model again to an 'agronomist–grader' model: an agronomist who will check on the produce and works as a grader who will check the quality of the produce (the agronomist also recruits and trains) during collection.

(mFarm Manager, 04/04/14)

In addition to taking a transaction fee, m-Farm has also developed a new role as a knowledge broker selling its data to research organizations looking at consumer behaviour and food scarcity. Furthermore, the expanding network was used to disseminate information relating to international regulations, for example, information about any pesticides that might be banned. Abass is now focused on the export market and has been in the UK to speak to large retailers who are keen to be more responsible in the way that they source their products. She states:

They want to have social responsibility . . . by sourcing produce through m-Farm they are playing a vital role in development and securing a consistent supply that is not dependent on middlemen.

(Solon, 2013)

10.5 Discussion and Conclusion

m-Farm's overriding objective was to achieve commercial sustainability whilst still retaining a social function. This required the scaling of m-Farm's operation with commensurate increases in income (both for m-Farm as an intermediary and the farmers as the final beneficiaries). To this end, the case study demonstrates a move from a market of 'middlemen' buyers and farmers who are individualized sellers, to collaborative forms of collective action. Collective action provides the means to fundamentally change farming from a fragmented and disorganized subsistence form to a more organized market-orientated approach which can provide the basis for the scaling of production, and potentially, step changes in agricultural productivity. In the case study, structural transformation goes hand-in-hand with the requirement for new forms of 're-intermediation' whereby m-Farm employs agronomists–graders as an alternative source of support to the Kenyan Ministry of Agriculture extension workers. Other m4Agric interventions such as the Grameen CKW Programme are also following this model by empowering community-based knowledge workers (who are farmers themselves) to act as 'info-mediaries'. These experiences are stimulating new approaches to agricultural extension, practiced through value-chain improvements and group-based action.

The redesign of value chains, both longitudinally along the value chain, and laterally via cross-functional networks, lies at the heart of the type of change driven by m4Agric services. In the case study, change takes place in relation to both the purchasing of input materials (aided by the group buying tool) and the selling of produce output (through collaborative redesign of produce marketing aided by the web-based and text-based tools for advertizing and completing transactions). Over four years since inception, m-Farm progressed from being purely a provider of information to becoming a trusted 'digital intermediary' that plays a central role in coordinating the value chain and completing transactions. This process of changing agricultural systems is complex, and requires considerable organizational effort as well as the necessary seed-corn financing. The case study benefited from initial and ongoing investment in digital infrastructure from donors, and as such, innovation expenses do not constitute fixed costs, they tend to increase over time, as applications need further development and eventual commercialisation to make them sustainable.

It seems as though services such as mFarm are coming to substitute for both the top-down role of the state (by providing an alternative to the traditional government extension services), and the bottom-up role of farmer collective action (traditionally organized through farmers' cooperatives). Within the confines of m-Farm's narrow client base and geographical spread, this change can be defined as transformational, in terms of a novel approach, the redesigned collaborative and marketing processes/relationships, expanded networks, both along the supply chain, and more broadly, reaching into and involving non-governmental and private stakeholders. It also could be suggested that this may be part of a wider pattern of power and influence accruing to new 'digital intermediaries' using digital development models.

Finally, understanding the characteristics of the produce sector market and how it operates is critical for successful ICT application. Interventions should build upon the specific characteristics of local demand, or the ability to identify specific farmers' or more importantly farmer groups' needs. Small-scale subsistence farmers may lack the incentives to collaborate or to grow because they remain embedded in a particular physical and institutional context, a context that mAgric initiatives may find difficult to transform. Local enablers of innovation processes are crucial for overcoming these obstacles, and in the development of successful mobile applications for agricultural development.

Notes

[1] See:https://www.gsma.com/mobilefordevelopment/m4d-tracker/magri-deployment-tracker (accessed 1 March 2018).
[2] See: http://techfortrade.org/ (accessed 1 March 2018).
[3] See: http://www.capitalfm.co.ke/business/2015/09/safaricom-opens-up-m-pesa-platform-to-developers/ (accessed 1 March 2018).

References

Burrell, J. and Oreglia, E. (2015) The myth of market price information: mobile phones and the application of economic knowledge to ICTD. *Economy and Society* 44(2), 271–292.

CKW (2016) Community Knowledge Worker Programme, Grameen Foundation. Available at: https://www.grameenfoundation.org/resource/lessons-learned-2009-2014-community-knowledge-worker-uganda-program (accessed 8 January 2018).

Duncombe, R.A. (2016) Mobile phones, agricultural and rural development: a literature review and future research directions. *European Journal of Development Research* 28, 213–235.

GSMA (2015) Market size and market opportunity for agricultural value-added services (AgriVAS), Market Analysis Report, GSMA. Available at: http://www.gsma.com/mobilefordevelopment/wp-content/uploads/2015/02/Agricultural-value-added-services-market-opportunity-and-emerging-business-models.pdf (accessed 17 December 2015).

Heeks, R. (2010) Do information and communication technologies (ICTs) contribute to development? *Journal of International Development* 22(5), 625–640.

Hildebrandt, N., Nyarko, Y., Romagnoli, G. and Soldani, E. (2014) Information is power? Impact of an SMS-based market information system on farmers in Ghana. CTED/New York University, New York, USA. Available at: https://editorialexpress.com/cgi-bin/conference/download.cgi?db_name=NASM2014&paper_id=856 (accessed 17 December 2015).

Poulton, C., Dorward, A. and Kydd, J. (2010) The future of small farms: new directions for services, institutions and intermediation. *World Development* 38(10), 1413–1428.

Qiang, C.Z., Kuek, S.C., Dymond, A. and Esselaar, S. (2011) Mobile applications for agriculture and rural development, ICT Sector Unit, The World Bank, Washington, DC, USA. Available at: http://siteresources.worldbank.org/INFORMATIONANDCOMMUNICATIONANDTECHNOLOGIES/Resources/MobileApplications_for_ARD.pdf (accessed 17/01/2014).

Solon, O. (2013) MFarm empowers Kenya's farmers with price transparency and market access. *Wired*, 21 June 2013. Available at: http://www.wired.co.uk/news/archive/2013-06/21/mfarm (accessed September 2017).

Stake, R.E. (2005) Qualitative case studies. In: Densin, N.K. and Lincoln, Y.S. (eds) *The Sage Handbook of Qualitative Research* (3rd edn). Sage, Thousand Oaks, California, USA, pp. 443–466.

Subervie, J. (2011) Evaluation of the impact of a Ghanaian mobile-based MIS on the first few users using a quasi-experimental design. Paper presented to the Workshop on African Market Information Systems, Bamako, 30 Nov–2 Dec. Available at: http://www.slideshare.net/Esoko/cirad-research-on-esoko (accessed 15 October 2016).

Van Campenhout, B. (2012) Mobile apps to deliver extension to remote areas: preliminary results from Mount Elgon area. Grameen Foundation. Available at: https://www.grameenfoundation.org/resource/mobile-applications-deliver-extension-remote-areas (accessed 15 December 2015).

World Development (2010) *Special Issue: The future of small farms* 38(10), 1341–1526.

World Bank (2011) *ICT in Agriculture: Connecting Smallholders to Knowledge, Networks and Institutions*. The World Bank, Washington, DC, USA. Available at: http://documents.worldbank.org/curated/en/455701468340165132/pdf/646050ESW0P1180lture0e0Sourcebook12.pdf (accessed 15 February 2014).

Yin, R.K. (2003) *Case Study Research: Design and Method* (3rd edn). Sage, Thousand Oaks, CA, USA.

11 Understanding the Impacts of Mobile Technology on Smallholder Agriculture

Stan Karanasios[1]* and Mira Slavova[2]

[1]RMIT University, Australia; [2]Gordon Institute of Business Science, University of Pretoria, South Africa

11.1 Introduction

The argument that digitization, computerization and increased information access through Information and Communication Technology (ICT) can improve lives, contribute to solving social problems and act as a vehicle for economic growth has gained currency in the early 21st century. Indeed, a growing number of reports by international development agencies and academic studies testify to an increased understanding of the linkages between ICTs and economic growth; for instance, studies show a relationship between high mobile penetration and economic growth (Waverman *et al.*, 2005; Cleeve, 2014).

With the arrival of mobile networks across the developing world, researchers, policy-makers and practitioners have looked for a tangible impact on agriculture. This has given rise to a growing body of research within the ICT for Development field (e.g., Prakash and De', 2007; Venkatesh and Sykes, 2013), as well as within the field of agricultural economics (e.g., Jensen, 2007; Aker *et al.*, 2016) on the direct welfare-enhancing impacts of mobiles in agriculture, with a particular focus on smallholder famers. This is consistent with the broader line of research which is concerned with how ICT can improve the livelihoods of people in the developing world (e.g., Barrett *et al.*, 2001; Heeks, 2010; Sahay *et al.*, 2010; Karanasios and Allen 2013).

Evidence points to mobiles being used amongst smallholder farmers for coordination, accessing market information, tracking financial transactions, interacting with agricultural experts and capturing information such as local market trends and recording agricultural demonstrations (Molony, 2008; Aker, 2010; Martin, 2011). While such studies are suggestive of patterns of use, recent research also finds that amongst smallholder farmers mobile technology remains underused, and the relevant agriculture content (e.g., how to manage pests, pricing, etc.) is plagued by issues such as scarcity and inaccuracy (Islam and Grönlund, 2011; Mubin *et al.*, 2015). At the same time, research continues to confirm the case in favour of legacy technologies (like radio) as cost-efficient and pervasive platforms for distribution and access to agriculture information and knowledge (Prakash and De' 2007; Venkatesh and Sykes 2013; Flor and Cisneros 2015).

One application of mobiles that has been prominent in the literature and in field interventions consists of their use for dissemination of market price information or 'market information systems'. Such systems carry the expected

---

* Corresponding author e-mail: stan.karanasios@rmit.edu.au

outcome of improving market efficiency and enhancing livelihood outcomes for smallholder farmers. Development aid agencies and private sector efforts aimed at the 'bottom-of-the-pyramid' have resulted in some of the most publicized services, such as Esoko (formerly TradeNet) in Ghana, and Reuters Market Light and Nokia Life Tools in India.

An early agricultural economics study by Jensen (2007) found that mobile network coverage is associated with improved prices for fishermen in India. It had the impact of nurturing the idea that cheaply delivered market price information (e.g., via SMS) was capable of rebalancing the status quo and enabling smallholder farmers and fishermen to improve their livelihoods. A key driver for this was the ability of smallholder farmers to circumvent intermediaries and seek new markets. This follows the central economic principle that for markets to function efficiently producers and traders require 'perfect' information, while inadequate information leads to inefficiencies and waste, both of which are common in African agricultural markets (Aker and Blumenstock 2015).

While much hype followed this study, subsequent studies have claimed less radical impacts and found more ambiguous evidence of the transformative impact of mobile market information. For instance, Aker (2010) and later Aker and Fafchamps (2015), found an association between mobile phones and small, but significant (between 6% and 16%), reduction in grain price dispersion. Others have noted increases in 'farm-gate' prices (Goyal 2010; Aker and Blumenstock 2015). Courtois and Subervie (2014) found that farmers in Ghana who benefited from market information services received significantly higher prices (by 7–10%) than they would have received had they not accessed mobile information. They suggest:

> that the theoretical conditions for successful farmer use of MIS (market information systems) may be met in the field.
> (Courtois and Subervie, 2014, p. 953)

Some economic studies have found little evidence of impact (Fafchamps and Minten 2012; Aker and Fafchamps 2015). An important finding is that while price increases may vary from marginal to more significant, evidence suggests that this represents return-on-investment of up to 200% (Hildebrandt et al., 2015).

Even though economists have offered partial explanations for these discrepancies due to market failures (Aker and Blumenstock, 2015), scholars descending from other disciplines in the social sciences, in particular information systems, have challenged the 'invisible hand' suggestion of improved market efficiencies due to mobile phone use. Most notably, recently Steyn (2016) critically re-examined Jensen's (2007) work and found no evidence of improved livelihoods, transformed market behaviours, nor absence of intermediaries; all supposedly resulting from mobile phone use. Providing a broader explanation Burrell and Oreglia (2013) explained inconsistencies across studies due to differences in normative rural practices and local challenges. They found that social factors such as cultural-historical relational norms carried more weight than price in the negotiation of market trades; and, information on pricing was not scarce and was available through reliable traditional rural (e.g., oral) communication channels. Taken together, this body of literature suggests that such factors play a strong role in how mobile initiatives are received, how they are used and how their impact on agriculture practices is realized.

Building on the perception that the impact of mobile use has a lot to do with localized behaviours and norms, we draw on a study of ICT use in rural Ghana, and in particular mobile technology (Slavova and Karanasios, 2018a, b). We show how mobiles introduce among smallholder farmers blended information practices that are consistent with both existing cultural-historical norms around farming (i.e., smallholder logic) and with policy imperatives aimed at re-casting farming 'as a business' and promoting value chain integration through ICT (i.e., 'value chain logic'). Additionally, we show how actors in rural agriculture – such as NGOs, broadcasters (e.g., local radio stations), SMS services (e.g., Esoko) and agriculture partners (e.g., Golden Stork) – are able to leverage such blended practices in advancing their value chain agenda. In designing interventions these actors put forward approaches, combining legacy technologies (e.g., radio), which align with oral tradition, with ones that correspond to contemporary business norms (e.g., mobile). We use these insights to deliver recommendations for information service providers, agriculture development practitioners and policy makers.

The remainder of this chapter is structured as follows. Section 2 frames the problem of competing logics facing smallholder farmers. Section 3 outlines the study conducted in rural Ghana between 2009 and 2014. We present the findings in Section 4. Section 5 discusses the findings and positions them within the broader literature. It also offers insights and recommendations for mobile use in agriculture development. Section 6 concludes the paper by summarising the key considerations.

11.2 Understanding Change in Smallholder Agriculture

When new market information technologies and services are introduced into rural settings, development practitioners, researchers and policy makers often predict impact on the basis of rational choice theory, whereby individuals make logical decisions based on costs and benefits. For instance, a common assumption can be that farmers will consider available pricing information and the probability of events (e.g. size of harvest) and undertake the self-determined best choice of action. Nonetheless, farmers' choices are usually guided by established relationships and norms. This includes cultural-historical norms around farming, interpersonal relations, strong ties with indigenous institutions and informal practices. Consequently, the size and speed of the impact of mobile information services depends on the institutional barriers thwarting changes in behaviour.

We suggest that often when mobile information services are introduced, the required behavioural change on behalf of farmers goes beyond learning how to use a new technology, or how to access a new service. Instead, beneficiaries are required to transform the historical patterns of their

> material practices, assumptions, values, beliefs, and rules by which individuals produce and reproduce their material substance, organize time and space, and provide meaning to their social reality.
>
> (Thornton and Ocasio, 1999, p. 804)

That is, the institutional logics that surround them. For example, when farmers – who cherish loyalty and strong personal ties (Burrell and Oreglia, 2013) – are expected to respond to newly available market price information by abandoning their established trading relationships, the change envisioned by development agents is at an institutional scale.

We suggest that the intended impact of mobile information services is often commensurate with a transition from organizing farmers' activities according to a smallholder logic, to organizing their activities according to a value chain logic. Such an impact cannot be pressure-cooked within the scope of short-term development programs lasting only a year or two, and it is rarely accounted for in monitoring and evaluation frameworks. Therefore, we propose that understanding of the impact of mobile, and subsequent information services, in terms of quantitative welfare improvements are largely misplaced. Instead, the impact should be viewed through the qualitative lens of changes in patterns of agricultural practices, and changes in assumptions. In other words, rather than limiting understandings of change to profits increases, change should be located within the actual practices of farmers. Fig. 11.1 summarizes the two logics within rural agriculture. The smallholder logic is consistent with agriculture as part of the normative rural life. The value chain logic is consistent with policy imperatives aimed at re-casting farming 'as a business' and promoting value chain integration through ICT (i.e., value chain logic).

We shift concerns with impact to understanding qualitatively farmers' actions within their institutional context, and interpreting the impact of ICT interventions in terms of change in institutionalized behaviours. This requires that we conceptualize mobile technology as existing within an ecology of complementary and substitutable technologies and practices. In this chapter, we address how ICTs, and mobile in particular, impact agriculture development; and how mobile can be used to foster agriculture development. We illuminate on the tensions and challenges, as well as opportunities, encountered in the move from traditional rural norms and cultural practices (i.e., smallholder logic), to ones based on market rationality and agronomic knowledge (value chain logic) (Slavova and Karanasios, 2014).

	Smallholder logic	**Value chain logic**
View of farming	Part of the rural 'way of life'	Agriculture as a business, top-down, policy driven
Relations	Interpersonal	Business contacts
Governance	Relational; indigenous institutions	Regulatory norms (e.g. measurement units, standards)
Trading practice	Through brokerage and intermediaries	Open market
Nature of practice	Unsophisticated; production with variable quality	Certifiable, knowledge and information intensive production
Role of mobile	To strengthen the smallholder logic	To enable a value chain logic

Fig. 11.1.   Smallholder and value chain logics.

11.3 The Study

Our study focused on three rural regions in Ghana: Northern Region; Ashanti; and Eastern Region. They are considered representative with considerable rural populations. We focused on eight communities, spanning different levels of infrastructure (e.g. surfaced roads, electricity) and a range of agricultural activities. Table 11.1 summarizes the study sites.

We combined data from different actors in the agricultural space and we drew on several data collection methods. This allowed us to go beyond studies whose understanding is built on a single data source. First, seven focus group discussions were undertaken with the help of local agriculture stakeholders across the three regions. The focus groups captured the perspectives from 119 farmers: Takorasi (13), Bonsaaso (18), Dawa (14), Kalande (16), Wudormeabra (22), Dalun (11) and Gushie (25). Each of the focus group discussions lasted approximately two hours and were carried out in the local language of the communities (with help from interpreters). The discussions focused on information channels and facilities, farming activities, ICT ownership and use.

Second, interviews were carried out with representatives of different rural agriculture actors offering information services. Organizations included local and international NGOs, radio stations, government departments, agriculture suppliers and technology start-ups. Interviews varied in length (45–82 minutes) and were conducted in English. They followed a semi-structured protocol and focused on their activities and information services and strategies using ICT and on exploring themes arising from the focus groups. Supporting the focus groups and interviews, we undertook observations of the work activities of farmers and information providers and visits to information facilities. We also reviewed other sources such as radio programs, institutional reports and learning briefs.

In addition, we draw on a national survey of Ghana from 2009 undertaken by *InterMedia*[1]. It is one of the most comprehensive data sources of ICT availability and information practices in Ghana. Using a standard questionnaire adapted to reflect the Ghanaian environment, *InterMedia* administered the survey in-person by approaching households. In total 2051 usable responses were obtained from 6720 contacts. For our purposes, we extracted a sub-sample of 305 households living in the selected regions (Ashanti 38.7%; Eastern 18.7%; and, Northern 42.6%), whose income in the past 12 months came

Table 11.1. Study sites.

Regions	Rural population	Total population	Mobile penetration	Internet users	Communities	Road access	Electricity
Ashanti	1,883,090	4,780,380	36.6%	2.3%	Takorasi	Yes	Yes
					Bonsaaso	No	No
Northern	1,728,749	2,479,461	11.9	0.7	Kalande	No	No
					Wudormeabra	No	No
					Dalun	No	Yes
					Gushie	Yes	No
Eastern	1,489,236	2,633,154	34.3%	2.2%	Dawa	Yes	Yes

predominantly from farming. The questions covered: (i) ICT availability and use; (ii) information needs and information practices; and (iii) trust. The survey followed a probability-proportional-to-size sampling plan based on the Ghana Statistical Service 2000 Population and Housing Census.

11.4 Findings: Mobile Technology and Changes in Smallholder Agriculture

In terms of access to ICT the survey data clearly identified radio (89.2%), mobile (63.6%), and TV (43.9%) as the technologies most farmers had access to. To a much lesser extent farmers had access to a computer (3.3%), landline (1.3%) and internet (1%). Drawing on our qualitative data we were able to interpret that some information sources and technologies for accessing information correspond to the smallholder logic, while others align with the value chain logic. The sources of information aligned with the smallholder logic are diverse and include in-person interactions and familiar print-and-broadcasting channels. We classify the following sources of information consistent with the smallholder logic in agriculture: traditional broadcasting (i.e. gong-gong beater, porter, town-crier, loudspeaker, etc.); radio and TV; newspapers and magazines, posters, billboards and brochures; and, in-person communication realized through social networks (i.e. family and friends, other farmers).

The radio had the most diverse uses. Farmers explained that weather information obtained

from the radio helps them make decisions about the application of fertilizers and it also helps them organize other operations on the farm (e.g. transplanting, harvesting, etc.). Other radio uses included learning about new products, learning how to use products, obtaining agricultural information, news and current events and public announcements. While radio falls under the umbrella of the smallholder logic it is considered mostly an individual media, and listening to the radio takes place within the family unit and the home. Occasionally groups would come together to listen to a programme and simultaneously discuss the topics.

Perhaps the only unexpected statistic was that TVs were relatively common. In the focus groups we found that despite poor access to the electricity grid, weak network signals and poor reception, TVs were not only common but also valued and aspired to. TV viewing had a social element and often several farmers would gather around one TV.

Meanwhile, sources consistent with the value chain logic include: ICT channels such as mobile SMS, and to a lesser extent the internet; and, formal organizational structures through which smallholder farmers access information (i.e. farmer organizations, cooperatives, and unions; extension offices; farming-supply vendors; and NGOs). While mobile is categorized under value chain logic, as we will describe it transcends both logics. Farmers often spontaneously brought up mobile technology as an important technology development that has been introduced in their communities over recent years. Most farmers spent between US$1 and US$3 a week on their phones. Access to the main electricity grid in rural Ghana was problematic, and several mechanisms for recharging mobile phones were established in rural communities – such as stand-by generators or at nearby friends, relatives, or neighbours.

Table 11.2 summarizes the propensity of farmers to seek agriculture-related information through different channels, based on the quantitative data. It paints a picture of Ghanaian farmers relying for their information needs extensively on channels consistent with the smallholder logic. Notably, print and broadcasting channels – and radio in particular – were the preferred sources of information about fertilizers (71.4%), soil problems (44.9%) and weather (57.4%). Meanwhile informal networks (56.4%) were the main source of market information. Such sources were complemented by channels carrying the value chain logic. In particular, the survey indicated that organized forms for accessing information (e.g., extension officers) were well recognized as a source of innovative information, such as information about new seed varieties (57.4%). As farming information was accessed fairly uniformly via formal channels, print-and-broadcasting, and informal channels, the data suggests that farmers combined channels symbolic of different institutional logics.

The internet was not reported as a significant source for agricultural information. While mobile SMS featured to a very limited extent, the survey data showed that 54.5% of farmers reported using their mobile phones in the last two days, and a further 7.4% used their mobile phones in the last week. However, it is important to note that the survey did not capture the use of mobile for voice interactions with family and friends, farmer organizations, cooperatives and unions. The survey included only mobile SMS (and not voice) as an optional information access channel. Furthermore, it did not collect data about the informal interactions carried out via mobile.

The qualitative data however, add broader meaning and interpretation to Table 11.2. The interviews and focus groups revealed a heavy reliance on mobile by farmers for interacting with peer farmers and farmer organizations, cooperatives, and unions as well as other sources of information. That is, mobile allows access to these other sources. The most common example was of farmers using the mobile phone to confirm information received through non-familiar sources with peers and friends. This was particularly common amongst farmers that received agronomic information via SMS. Even radio (typically consistent with the smallholder logic) was strengthened by mobile by allowing farmers to call in to ask questions to experts (interacting

Table 11.2. Access to agricultural information (respondents could give more than one answer) ($n=305$).

Logic	Information source	Market prices (%)	Fertilizers (%)	Seed varieties (%)	Soil problems (%)	Weather (%)
VALUE CHAIN LOGIC	Mobile SMS	0	0	0.3	0	0
	Internet	0	0	0	0	0
	Farmer organizations, cooperatives and unions	10.8	14.7	12.5	11.8	11.8
	Extension office	20.3	29.2	32.1	25.2	23.6
	Farming-supply vendors	8.9	9.8	8.5	3.6	3.6
	NGOs	2.6	3.6	4.3	3.3	2.6
SMALLHOLDER LOGIC	Radio	32.1	46.2	37.7	29.8	38
	TV	13.1	23.9	13.8	13.8	17.7
	Newspapers, magazines	0.7	2	0.3	0.3	0.7
	Posters, billboards, brochures	0.3	2	0.7	1	1
	Family, friends	27.5	26.2	22.3	18.7	19
	Other farmers	28.9	28.5	27.2	23.6	23.6

Source: Adapted from CIA (2014) and GSS (2012).

with experts is typically consistent with the value chain logic). Radio broadcasters also leveraged mobile technology and integrated voice response (IVR). IVR blends the novelty of new technology and accessing information when convenient, with a traditional preference for oral communication. It was used to send an SMS alert to farmers on programme schedules, announce telephone numbers of extension agents online and for farmers to call and listen to recorded programmes. In other words, the use of mobile transcended both logics and can be viewed as an enabler of blended information practices.

Other common uses emerging from the qualitative data include 'flashing' (whereby a caller would dial a phone number in order to receive a call back and is an acute example of how mobile technology affords the enhancement of existing information channels), timekeeping and taking pictures. The use of mobile applications was non-existent.

The findings surrounding how the use of different information sources are used for agriculture and how mobile is used to connect farmers with different sources of information are supported by Table 11.3, which summarizes the level of trust in each source. It clearly demonstrates that trust is significantly higher amongst the sources in line with the smallholder logic as well as the extension office and farming supply vendors (as denoted by the shaded boxes). This shows that farmers value relational networks and interpersonal relations. It also suggests that

mobile initiatives need to be built around these norms.

The finding of blending institutional logics of ICT use among farmers can be translated into a finding of mixed impact approaches by agriculture development stakeholders. For instance, a mobile information service provider used SMS to introduce new market price information in standard metric weight measurements (e.g. kilograms), rather than traditional volume units (e.g. bags, bowls etc.). This service had two dimensions: (i) novel information (i.e. kilograms rather than traditional units and price information); and (ii) novel delivery method (i.e. by mobile SMS). Because of the novelty of both the information and the communication medium, the challenges faced by smallholder farmers in adopting the service were twofold. Accustomed to traditional volume units, they could not make sense of, and estimate the value of their produce using prices denominated in kilogram. Additionally, they had difficulties accessing and reading information sent via the mobile SMS channel. The development actors responded to this tension by engaging fieldworkers who could explain to smallholder farmers the meaning of weight measurements and of the received SMS. Furthermore, fieldworkers demonstrated to farmers how to use SMS on their phones. This meant that the service providers' impact approach relied on mobile, alongside fieldworkers who generated trust and understanding of the new service.

Table 11.3. Trust towards agricultural information sources (respondents could give more than one answer) (*n*=305).

			Very untrusted	Somewhat untrusted	Don't know	Somewhat trusted	Very trusted
VALUE CHAIN LOGIC		Mobile SMS	1.3	3.6	79	7.9	7.9
		Internet	0.7	1.3	87.5	4.6	5.6
		Extension office	0.7	3.3	41	17.4	37.7
		Faming supply vendors	1.3	7.5	32.1	36.1	23
SMALLHOLDER LOGIC		Radio	0	0	21.3	21	57.7
		TV	0.3	0.3	44.3	15.7	39.3
		Newspapers, magazines	1.3	2	76.7	8.5	11.1
		Posters, billboards, brochures	1.6	2	74.4	7.5	14.1
		Family, friends	1.3	2.3	22	32.8	41.6
		Other farmers	1.6	4.6	21	33.1	39.7

Cases of non-novel information and novel dissemination methods are also indicative of mixed impact approaches. For instance, an NGO developed the Talking Book (a handheld audio device which had pre-recorded information; farmers could also record their own voices[2]), a novel and unfamiliar technology which delivered, among other information, advice on germination testing on the recycled seeds used by farmers. Rather than promote new seeds, which is considered a radical change by farmers; the Talking Book approach was to emphasize the importance of testing existing seeds, which is closer to farmers' current practices. The Talking Book was physically passed on from farmer to farmer. It was especially relevant for illiterate farmers and those in very remote areas where mobile and radio was challenging. Farmers who did not have access to Talking Book devices visited farmers in nearby villages. By using the novel delivery mechanism to provide information that did not challenge the logic of smallholders' practices, the Talking Book was able to generate a step change.

Broadcasters' also adopted mixed impact approaches by blending mobile with radio. SMS campaigns and alerts and IVR systems were used to increase smallholder farmers' exposure to agricultural information and their awareness of improved practices; all novel practices in the context of rural Ghana. SMS campaigns and alerts involved broadcasting extension agent's telephone numbers as well as reminders on when relevant content would be broadcast. This balanced the trust associated with familiar radio broadcasts with the tension surrounding value chain content (e.g. extension agents/traders phone numbers).

These examples serve to illustrate the mixed impact approaches pursued by agriculture development stakeholders. Table 11.4 provides a summary of those approaches, specifically highlighting mobile technology and agronomic information.

Table 11.4 also shows two combinations relevant to mobile. First, where the agronomic message is unfamiliar and contests existing practices (e.g. market price information, promotion of new seed varieties) then it is necessary to complement mobile with information channels and sources consistent with the smallholder logic. Second, where the agronomic message is not new or contested then the risk is much lower.

These mixed approaches were learnt over time. During the course of our study we observed how the market information service Esoko adapted their approach from mobile centric and high impact to a mixed approach. At the start of our study, through their platform they offered an SMS market price service. The intended benefit of the service consisted of enabling direct market access for farmers and cutting off intermediaries. The adoption of the SMS service by farmers was plagued by problems of understanding metric measurements, usability of the SMS technology and rural illiteracy. By the end of our study, the approach of Esoko was much more mixed. Esoko was complementing the SMS service with a call centre which offered access to market prices through the voice channel. Farmers

Table 11.4. Development stakeholders' use of mixed impact approaches to mobile.

Agronomic message contests current practice	Mobile phone delivery	Interventions of development stakeholders
Yes	Yes	High-risk/high-impact approach: in addition to mobile, agriculture actors must ensure that the agronomic message (e.g. market price information) is also delivered and supported via familiar channels (e.g. informal face-to-face, radio)
Yes	No	Low risk approach: contested agronomic message is delivered via familiar channels (e.g. informal face-to-face, radio)
No	Yes	Low risk approach: use mobile to deliver non-novel information. Because the information is not unfamiliar, additional sense-making is not needed by farmers to support mobile (e.g. radio broadcaster's use of IVR and SMS alerts)

responded with high levels of trust to market prices received from call centre operators during personalized calls. Meanwhile, the intended impact of the service was moderated. Instead of expecting farmers to benefit from market switching behaviours, the Esoko service was geared towards improving transparency in agriculture marketing and forging trust among farmers and traders.

11.5 Discussion

The premise of this chapter is that ICTs – particularly mobile – impact agriculture development by facilitating step-wise changes in farmers' information and agriculture practices. The transition of farmers from operating under the smallholder logic to operating under the value chain logic is enacted through the intermediate stage of 'hybrid logic' which blends elements of information use and agriculture practices characteristic of either logic. In this chapter, we focused on the role of mobile in facilitating the emergence of the hybrid logic through the activities of smallholder farmers and development actors. We emphasize the focus on step change as opposed to transformation, which would be conceptualized as farmers switching their practices to ones consistent with the value chain logic. Our argument for step change is supported by recent studies which have measured the economic impacts of SMS market price systems by smallholder farmers (Aker and Blumenstock, 2015; Courtois and Subervie, 2014; Goyal,

2010). Fig. 11.2 illustrates the hypothetical institutional process and the role of mobile. While the blending of the two logics is located equally distant between the smallholder and the value chain logics, it is not to suggest that it is at a midway point of an irreversible process of change. Rather, Fig. 11.2 summarizes our finding that current practices draw on both logics. This perspective rejects the notion that simply providing farmers with information or an ICT leads to transformation, but rather that change comes from understanding how information and information tools may shift institutionalized practices. That is, examining changes in practices (e.g. blending of logics) rather than simply access to information services is likely to deliver more meaningful understanding of change. It also follows that change is only likely to occur if underlying structural constraints, poor and inefficient practices and market failures are addressed (Aker and Blumenstock, 2015).

Considering prescriptive insights as to how ICTs should be used in order to foster agriculture development (i.e. the shift of institutional logics in agriculture), we find that combining elements aligned with different institutional logics into a single intervention is a promising approach. In particular, for ICT interventions in the agriculture sector, it is advisable to develop impact strategies blending information access mechanisms characteristic of the smallholder logic (e.g. radio, social networks, etc.) with ICTs characteristic of the value chain logic (e.g. mobile). Such mixed strategies constitute a viable approach for development agencies seeking to advance the logic of the agriculture sector.

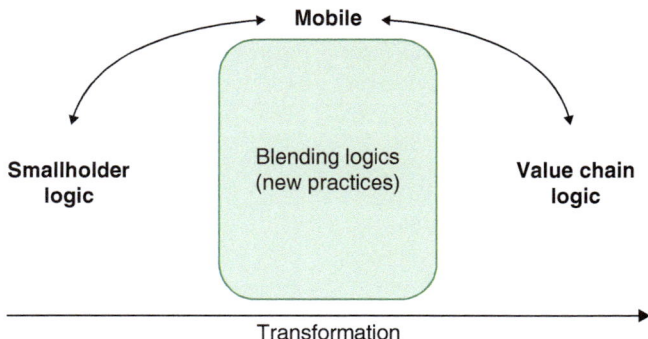

Fig. 11.2. Blending logics through mobile.

11.5.1 Recommendations for researchers and practitioners

Our research identified several recommendations for researchers and practitioners concerned with mobile technology and changes in smallholder farmer practices. These are:

1. Our study clearly highlights how mobile technology complements, rather than substitutes, existing information sources. Most studies tend to pinpoint mobile solutions only and neglect the broader information ecology. Information service providers need to understand the interplay between different technologies and non-technology information sources to develop successful and contextually relevant services.

2. In early prototypes, practitioners need to leverage the versatile affordances of mobile by allowing for all possible interactions – voice and text, one-way as well as two-way. Consider the reasons for eliminating any of the possibilities.

3. Always consider the novelty of the content, information, interaction, practice or exchange that is being introduced. Offset novelty by: (i) relying on trusted intermediaries and popular communication formats (e.g. oral); and (ii) enabling people to ask questions, i.e. discussion/two-way. Such formats are able to translate new knowledge and practices into understandable behaviour changes.

4. Determine what the goal of the information service is. For example, is its goal limited only to providing access to market price information? Or is its goal broader such as to encourage changes in practice (e.g. where and how farmers sell products, which inputs they use, etc.)? The former goal can be achieved through technology means only and can be measured in short-term impact assessments. While improving access to information is consistent with the value chain logic, the real impact is often minimal. Achieving broader impact goals however, requires consideration of practices at the institutional level. Embedding new information within the framing of cultural-historical norms and triggering blended practices is more likely to lead to substantial, albeit incremental change.

11.5.2 Future considerations

While our research focused on mobile technology and smallholder farmers we identify two major broader, yet related, considerations for future research and practice on ICT and agriculture in Africa. The first concerns the challenges around advances in big data use in agriculture for precision agriculture and supporting decision making – much of which will be facilitated by mobile. Traditional approaches to farming and indigenous knowledge are coming into tension with advisory technologies, geared towards generating operating efficiencies (e.g. reducing input costs, negating environmental impact, cutting down fatigue) and optimizing yields. The leading assumption behind such technologies consists of the far-reaching power of scientific measurements, derived via sensors, mobile data and devices, drones, or other remote equipment-mounted hardware. Once collected, such data are processed by software-enabled control systems, yielding advisory output which results into real-time field activities and adjustments of equipment. Such overreliance on scientific methods comes into conflict with understandings of agriculture as a situated social practice, bound within rural worldviews and identities. While the rigor and benefits of precision agriculture technologies are difficult to dispute, our research suggests it is worth pointing out that achieving their potential is contingent on aligning their outputs with farmers' existing view of agriculture and with their ongoing ICT use practices. When farming is viewed as an ancestral practice, developed through communal knowledge sharing; taking full advantage of systems delivering individually-targeted, real-time advisories may require social infrastructures that cannot be easily superimposed. The impact of such systems may be contingent on re-casting agriculture not only as a business but also as a scientific practice; rather than as an ancestral cultural practice. Alternatively, designers of precision agriculture technologies may be open to moulding them according to the existing cultural landscape. They might be willing to leverage traditional oral cultures and knowledge sharing mechanisms.

The second challenge concerns expanding the frame of reference of mobile information services to account for concerns around climate change and the subsequent projected increases in the frequency and intensity of extreme weather events and droughts and water shortages (UNISDR, 2015). This will lead to the need for new inputs, changes in practices and information needs and reliance on ICT (Karanasios, 2011; Pettengell, 2015). Lessons need to be learnt on how current mobile information services change patterns of behaviour, what works and what doesn't in order to transpose these lessons.

11.6 Conclusion

This chapter presented the case of how ICTs – particularly mobile – impact agriculture development by facilitating step-wise changes in smallholder farmers' ICT use and agriculture practices in Ghana. We showed how mobiles introduce among smallholder farmers blened information practices that are consistent with: (i) cultural-historical norms around farming and ICT use (i.e. smallholder logic); and (ii) policy imperatives aimed at re-casting farming 'as a business' and promoting value chain integration through ICT (i.e. value chain logic). We emphasize the focus on step change to a hybrid logic as opposed to transformation, which would be conceptualized as farmers transitioning to a value chain logic. We used these insights to proide recommendations for information service providers, agriculture development practitioners and policy makers on how to design and deliver information serices to smallholder farmers.

Acknowledgments

We thank InterMedia for access to the Audienc-eScapes Ghana survey data. We also acknowledge the support for this research by the International Food Policy Research Institute (IFPRI), Washington, and the University of Leeds and RMIT University for its support. We are gratefully recognize the time and support of the research participants and facilitating organizations. Finally, we are appreciative of the reviewers and editor, whose comments and thoughtful suggestions improved this chapter.

Notes

[1] See: http://www.intermedia.org/research-findings/audiencescapes/ (accessed 1 March 2018).
[2] See: http://www.literacybridge.org/talking-book/ (accessed 1 March 2018).

References

Aker, J.C. (2010) Information from markets near and far: mobile phones and agricultural markets in Niger. *American Economic Journal: Applied Economics* 2(3), 46–59.

Aker, J.C. and Blumenstock, J.E. (2015) The economic impacts of new technologies in Africa. In: Monga, C. and Lin, J.Y. (eds) *The Oxford Handbook of Africa and Economics: Volume 2: Policies and Practices.* Oxford University Press, Oxford, UK.

Aker, J.C. and Fafchamps, M. (2015) Mobile phone coverage and producer markets: evidence from West Africa. *World Bank Economic Review* 29(2), 262–292.

Aker, J.C., Ghosh, I. and Burrell, J. (2016) The promise (and pitfalls) of ICT for agriculture initiatives. *Agricultural Economics* 47(S1), 35–48.

Barrett, M., Sahay, S. and Walsham, G. (2001) Information technology and social transformation: GIS for forestry management in India. *Information Society* 17(1), 5–20.

Burrell, J. and Oreglia, E. (2013) The myth of market price information: mobile phones and epistemology in ICTD. *Working Paper,* University of California, Berkeley, California, USA.

CIA (2014) *The World Fact Book.* Available at: https://www.cia.gov/library/publications/cia-maps-publications/map-downloads/ghana_admin.jpg/image.jpg (accessed 12 March 2014).

Cleeve, E. (2014) Mobile telephony and economic growth in Africa. *Thunderbird International Business Review* 56(6), 547–562.

Courtois, P., and Subervie, J. (2014) Farmer bargaining power and market information services. *American Journal of Agricultural Economics* 97(3), 953–977.

Fafchamps, M. and Minten, B. (2012) Impact of SMS-based agricultural information on Indian farmers. *World Bank Economic Review* 26(3), 383–414.

Flor, A.G. and Cisneros, A.J. (2015) e-Agriculture. In: Mansell, R. and Ang, P. H. (eds) *International Encyclopedia of Digital Communication and Society*. John Wiley, Chichester, UK.

Goyal, A. (2010) Information, direct access to farmers, and rural market performance in Central India. *American Economic Journal: Applied Economics* 2(2), 22–45.

GSS (2012) Population and Housing Census. Accra: Ghana Statistical Service.

Heeks, R. (2010) Do information and communication technologies (ICTs) contribute to development? *Journal of International Development* 22(5), 625–640.

Hildebrandt, N., Nyarko, Y., Romagnoli, G. and Soldani, E. (2015) Price information, inter-village networks, and 'bargaining spillovers': experimental evidence from Ghana. Available at: https://editorialexpress.com/cgi-bin/conference/download.cgi?db_name=CSAE2015&paper_id=1059 (accessed 16 January 2016).

Islam, M.S. and Grönlund, Å. (2011) Bangladesh calling: farmers' technology use practices as a driver for development. *Information Technology for Development* 17(2), 95–111.

Jensen, R. (2007) The digital provide: information (technology), market performance, and welfare in the South Indian fisheries sector. *The Quarterly Journal of Economics* 122(3), 879–924.

Karanasios, S. (2011) *New and emergent ICTs and climate change in developing countries*. Centre for Development Informatics, University of Manchester, Manchester, UK, 39pp.

Karanasios, S. and Allen, D. (2013) ICT for development in the context of the closure of Chernobyl nuclear power plant: an activity theory perspective. *Information Systems Journal* 23(4), 287–306.

Martin, B.L. (2011) Mobile phones and rural livelihoods: diffusion, uses, and perceived impacts among farmers in rural Uganda. *Information Technologies and International Development* 7(4), 17–34.

Molony, T. (2008) Running out of credit: the limitations of mobile telephony in a Tanzanian agricultural marketing system. *Journal of Modern African Studies* 46(4), 637–658.

Mubin, O., Tubb, J., Novoa, M., Naseem, M. and Razaq, S. (2015) *Understanding the needs of Pakistani farmers and the prospects of an ICT intervention*. Paper presented at the 33rd Annual ACM Conference Extended Abstracts on Human Factors in Computing Systems, Seoul, Republic of Korea.

Pettengell, C. (2015) *Africa's Smallholders Adapting to Climate Change*. Oxfam International, Oxford, UK.

Prakash, A. and De', R. (2007) Importance of development context in ICT4D projects: a study of computerization of land records in India. *Information Technology and People* 20(3), 262–281.

Sahay, S., Sæbø, J.I., Mekonnen, S.M. and Gizaw, A.A. (2010) Interplay of institutional logics and implications for deinstitutionalization: case study of HMIS implementation in Tajikistan. *Information Technologies and International Development* 6(3), 19–32.

Slavova, M. and Karanasios, S. (2014) *Legitimacy of agriculture extension services: understanding decoupled activities in rural Ghana*. Paper presented at the European Group on Organization Studies, 30th EGOS Colloquium, Rotterdam, The Netherlands.

Slavova, M. and Karanasios, S. (2018a) When institutional logics meet ICTs: examining hybrid information practices in Ghanaian agriculture. *Journal of the Association for Information Systems* (in press).

Karanasios, S. and Slavova, M. (2018b) How do development actors 'do' ICTD: a strategy as practice perspective. *Information Systems Journal* (in press).

Steyn, J. (2016) A critique of the claims about mobile phones and Kerala fisherman: the importance of the context of complex social systems. *Electronic Journal of Information Systems in Developing Countries* 74(3), 1–31.

Thornton, P.H. and Ocasio, W. (1999) Institutional logics and the historical contingency of power in organizations: executive succession in the higher education publishing industry, 1958–1990. *American Journal of Sociology* 105(3), 801–843.

UNISDR (2015) *The Human Cost of Weather-Related Disasters 1995–2015*. Centre for Research on the Epidemiology of Disasters, United Nations Office for Disaster Risk Reduction, Geneva, Switzerland.

Venkatesh, V. and Sykes, T.A. (2013) Digital divide initiative success in developing countries: a longitudinal field study in a village in India. *Information Systems Research* 24(2), 239–260.

Waverman, L., Meschi, M. and Fuss, M. (2005) The impact of telecoms on economic growth in developing countries. In: *Africa: the Impact of Mobile Phones*. Vodafone Policy Paper Series, Number 2, London, UK, pp. 10–23.

12 Farmerline: A For-profit Agtech Company with a Social Mission

Worlali Senyo*

Farmerline, Accra, Ghana

12.1 Introduction

Farmerline has found a space for innovation and profit through technology in a massive agriculture market serving billions of people in the developing world. Food production in developing countries needs to almost double if the world is to feed its population of 9.1 billion people by 2050 (FAO, 2009a). Spreading information, knowledge and technology for improved food production by agricultural extension agents is a major challenge (Ragasa *et al.*, 2016; Tata and McNamara, 2016). In sub-Saharan Africa, the lack of funding for travel and regular training, shortage of field agents, poor roads and communication infrastructure has created a huge gap in market access and opportunity for farmers (FAO, 2009b; FAO, 2009c). With rapid pace of technological advancements and innovations, Information and Communication Technologies (ICTs) have shown promising evidence for providing solutions and tools for information dissemination and collection at relatively cheaper cost while overcoming the many barriers to traditional extension approaches. Despite these opportunities, ICTs (particularly mobile technologies) require innovative and financially sustainable business approaches to delivering extension and support services that are relevant, accessible and affordable to smallholder farmers (OECD, 2010; Steyn *et al.*, 2013).

12.2 Farmerline Evolution

The idea to start the company came out of the Mobile Web Ghana's Apps Competition in 2011 organized by the World Wide Web Foundation[1] where the two co-founders, Alloysius Attah and Emmanuel Owusu Addai, paired to develop a solution to send farming tips to farmers via text messages. Their application was selected winner of the competition and won a prize of US$600 which became the start-up capital for the company (Froumentin, 2012). The company, after running a pilot for a few months, realized that text messages were not providing the impact that was anticipated because a significant segment of the target farmers could not read and write. The team adapted the solution to include voice messaging based on feedback from farmers and extension officers, which enabled less-literate farmers access to the information services. Farmerline has since been deploying the technology in Ghana's agricultural sector.

Farmerline has built a social business, software platform and partnership networks across five countries through agribusiness, NGOs and technology partners over the past four years and is seeking to transform millions of smallholder farmers into successful entrepreneurs. By leveraging the local knowledge and skills of the agricultural sector, Farmerline has become a player on the international stage, offering competitive

---

* E-mail: worlali@farmerline.org

and innovative solutions for companies who need the local expertise that Farmerline can provide. Farmerline bases all of its technology on a close relationship with the farmers they serve, starting at home in Ghana. Farmerline can find added-value across stakeholders who may look very different at first glance – and it turns out that big European corporations and a 65-year-old cocoa farmer in Ghana or Sierra Leone actually need many of the same tools to reach their goals. Acting as an infomediary who puts farmers first, understands big business, and seeks these mutually beneficial situations, Farmerline has found a sweet spot of innovation in ICT – harnessing the power of native expertise and building an international company out of it.

12.3 ICT Tools and Business Model

Farmerline's primary ICT tool is its proprietary MERGDATA software. MERGDATA disseminates information such as weather forecasts, market prices, financial literacy and agronomic tips to farmers through mobile technology adapted with voice messages in any local language to improve farmer productivity (Anastasios *et al.*, 2010; Oladele, 2011) (*see* Fig. 12.1). It also offers an efficient data collection and analytics

feature allowing value chain actors to survey and profile farmers, map farm plots, manage supply chains with inventory tracking and cataloguing, and facilitate payments to farmers via mobile money (*see* Fig. 12.2). The in-house team of world-class programmers built MERGDATA, with the team hand-selected from the top computer science programmers in the country. These programmers choose Farmerline for its strong roots to service and people while reaching for one of the most innovative for-profit models the agriculture space in West Africa has seen. The combination of excellence and creativity in programming alongside dedication to mission has created a technology that is intuitive, powerful, and easy to sell.

Farmerline has a two-pronged profitable business model driven by Business-to-Business (B2B) revenues through licensing of its software-as-a-service (SaaS) to small, medium and large organizations and associations who work with networks of farmers, and through direct sale of agricultural and nutritional content services to farmers in their local language – Business-to-Farmer (B2F) sales – outside the reach of B2B partners (*see* Fig. 12.2). The bifurcation of this model allows for ultimate engagement across the value chain. As of 2015, the company's financial position showed a strong and profitable business outlook. Currently the company

Fig. 12.1. Screenshot of MERGDATA messaging application dashboard.

has reached over 200,000 users across West Africa and plans to scale further to 1 million active users by 2020 through technology and business partnerships across Africa. For truly revolutionary ICT and agriculture products, Farmerline has to stay close to farmers while also cultivating relationships with the multinational companies, smaller agribusinesses, and other supporting organizations such as Mobile Network Operators (MNOs), NGOs and governmental institutions that influence the space. Beyond profits, the social mission of the company is central to its very existence which Dr MirHossein Tabatabaei notes:

> Giving back is not only volunteering or raising money for food . . . giving back to society is to also support local entrepreneurs to independently, proactively, and sustainably produce value for themselves and their local communities.

(Tobias, 2014, p. 1)

Farmerline in ensuring its technology achieves the desired impact in the lives of farmers, is building on its structure to ensure a more systematic approach to measuring and reporting farming activities. Most evidence recorded has been on the level of small case studies or pilot studies.[2] Farmerline's strategy has always been to understand the value chain and the space around it, and look to address the key pain points for its customers. Through our direct sales to farmers, Farmerline gains a stronger understanding of the needs of farmers, their willingness and ability to pay, brand connection, and touchstone for ensuring the technology provided matches the rate of adoption in the communities we targeting.

12.4 Conclusion

Farmerline's success is built on the deep connection to the problems they aim to solve and the commitment to innovate in a unique way that is contextualized, lean and agile (Blank 2013; Ries, 2011). MERGDATA is a powerful tool, but the way that the tool is used is equally as important to the success of the company as the quality of the software. The Farmerline model provides direction for other companies in the ICT space to start from the wealth of local expertise at the grassroots level when approaching development problems. It further shows that the for-profit model, when taken as a framework for innovation, can provide the infrastructure for empowerment of customers and clients, but also for those that have built the company. Agriculture provides a particularly inspiring backdrop for this success. In order to address the vast need for food and livelihoods around

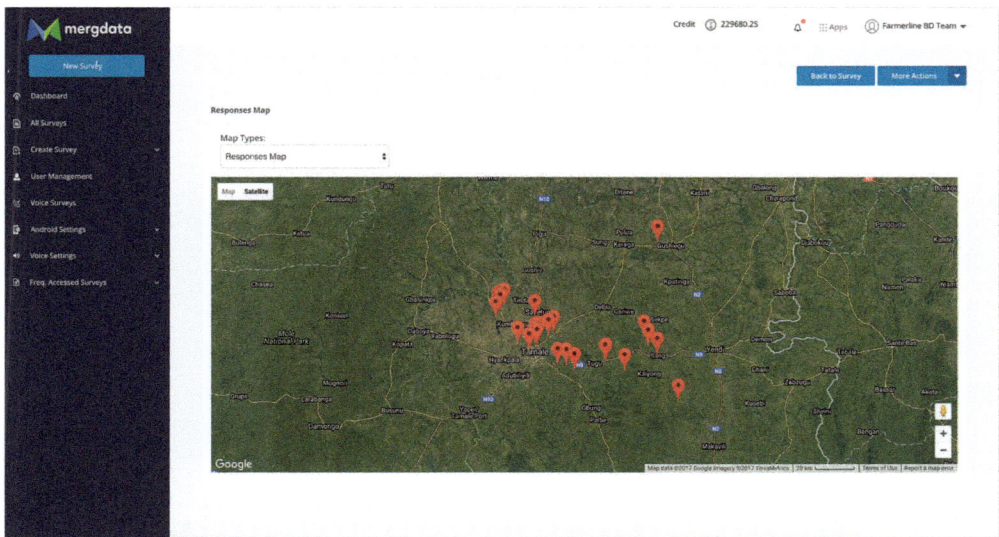

Fig. 12.2. Screenshot of MERGDATA survey responses map visualization.

the world, and to push some of the massive worldwide market towards the smallholders, there is an exceptional potential for re-imagining the rules of engagement. Farmerline is eager to see others re-imagine the ICT and agriculture sector by starting with their local expertise, to see a new wave of grassroots innovation taking hold.

Notes

[1] The World Wide Web Foundation was established in 2009 by Web inventor Sir Tim Berners-Lee to advance the open Web as a public good and a basic right. Read more at http://webfoundation.org/ (accessed 1 March 2018).

[2] Refer to Haggard, S. (ed.) (2015) and O'Reilly, M. (2014) for more information on the pilot case studies.

References

Anastasios, M., Koutsouris, A. and Konstadinos, M. (2010) Information and communication technologies as agricultural extension tools: a survey among farmers in West Macedonia, Greece. *Journal of Agricultural Education and Extension* 16(3), 249–263.

Blank, S. (2013) Why the lean start-up changes everything. *Harvard Business Review* 91(5), 63–72.

FAO (2009a) *High level expert forum – how to feed the world in 2050: global agriculture towards 2050.* Agricultural Development Economics Division: Economic and Social Development Department, Food and Agriculture Organization, Rome, Italy.

FAO (2009b) *High level expert forum – how to feed the world in 2050: the special challenge for sub-Saharan Africa.* Agricultural Development Economics Division: Economic and Social Development Department, Food and Agriculture Organization, Rome, Italy.

FAO (2009c) *High level expert forum – how to feed the world in 2050: the technology challenge.* Agricultural Development Economics Division: Economic and Social Development Department, Food and Agriculture Organization, Rome, Italy.

Froumentin, M. (2012) Farmerline wins mobile web Ghana's app competition. *Webfoundation.org* News and Blog, 22 March. Available at: http://webfoundation.org/2012/03/farmerline-wins-mobile-web-ghanas-app-competition/ (accessed 10 January 2017).

Haggard, S. (2015) Farmerline (Ghana). In: Zeng, H. *et al.* (eds) *Approach of ICT in Education for Rural Development: Good Practices from Developing Countries.* Sage, New Delhi: India, pp. 528–544. Available at: https://books.google.com.gh/books?id=aRqJCwAAQBAJ (accessed 21 April 2017).

OECD (2010) *SMEs, Entrepreneurship and Innovation.* OECD Publishing, Paris.

Oladele, O.I. (2011) The effect of information and communication technology on agricultural information access among researchers, extension agents, and farmers in South Western Nigeria. *Journal of Agricultural Education and Extension* 12(2), 167–176.

O'Reilly, M. (2014) Guest post from Farmerline: Ghan'a Fishin'. Available at: https://indigotrust.org.uk/2014/08/26/guest-post-from-farmerline-ghana-fishin/ (accessed 21 April 2017).

Ragasa, C., Ulimwengu, J. and Randriamam, J. (2016) Factors affecting performance of agricultural extension: evidence from Democratic Republic of Congo. *The Journal of Agricultural Education and Extension* 22(2), 113–143.

Ries, E. (2011) *The Lean Startup: How Today's Entrepreneurs Use Continuous Innovation to Create Radically Successful Businesses.* Crown Publishing, New York, USA.

Steyn, J., Rampa, M. and Marais, M. (2013) Participatory development of ICT entrepreneurship in an informal settlement in South Africa. *The Journal of Community Informatics* 9(4). Available at: http://ci-journal.net/index.php/ciej/article/view/969/1057 (accessed September 2017).

Tata, J. and McNamara, P. (2016) Social factors that influence use of ICT in agricultural extension in Southern Africa. *Agriculture* 6(2), 15.

Tobias, H. (2014) Paving the way for a better world: DeGroote Ph.D. student, MirHossein Tabatabaei, travels to Ghana. Available at: http://www.degroote.mcmaster.ca/articles/paving-way-better-world-degroote-ph-d-student-mirhossein-tabatabaei-travels-ghana/ (accessed 20 January 2017).

13 Best Practice Lessons and Sources of Further Information

Richard Duncombe*

Centre for Development Informatics, University of Manchester, UK

As the cases assembled in this book demonstrate, smallholders are typical of agriculture in the Global South. For this sector, improved agricultural productivity and greater overall efficiency are required to move to more modern – and probably larger-scale – forms of farming, moving on from predominantly subsistence farming to an expanded role for market-oriented production. New knowledge and capabilities combined with the effective delivery of rural services is seen as '... essential if small farms in high potential areas are to intensify production, contribute to economic growth and reduce poverty ...' (Poulton *et al.*, 2010, p. 1413). As the previous chapters demonstrate, digital technologies can play an important role in the Global South, supporting a broad range of interventions which are designed to strengthen research and innovation in the agricultural sector.

This final chapter summarizes the best practice lessons that have arisen from the experience and evidence presented. The best practice lessons are based only on the cases that have been included in this book. However, given that the cases include the design and implementation of a broad range of initiatives from differing country contexts, we would hope that the lessons learned are equally broad and comprehensive. It should be emphasized, however, that the focus of the book is principally at the micro-level of analysis where practitioners are working to implement projects and programmes on the ground,

and does not take account of the broader institutional and policy environment, neither for agricultural development nor for digital technologies, which was not a focus of this book.

13.1 Best Practice Lessons

13.1.1 Effective community engagement

- Implementing digital technologies in the agricultural settings of developing countries is 'as much about people as it is about technology'.
- The starting point for any intervention involving new technologies should be effective community engagement. The cases set out in the book provide evidence of this and a number of different approaches are suggested.
- Interventions should work from best practice (e.g. 'technology stewardship' in Chapter 5) to develop community engagement, to plan, design, and implement technology prototypes which involve establishing user needs and understanding location-specific constraints.
- A detailed understanding of the characteristics of the produce sector and how it operates is critical for successful application of digital technologies.

---

* E-mail: richard.duncombe@manchester.ac.uk

- Interventions should build upon the specific characteristics of local demand, or the ability to identify specific farmers' or more importantly farmer groups' needs.
- Local enablers of innovation processes are crucial in the development of successful mobile applications and the evidence suggests that achievement of scale through collective action is an effective and productive way to transform pre-existing farming systems.

13.1.2 Quality and usability of content

- 'Data, information and knowledge' lie at the heart of digital technologies and systems, and issues of information quality should be prioritized over the technological means to communicate that information.
- Raw data have to be processed to create usable information that farmers can use, and information should be communicated in a form that can be assimilated effectively by users to support decision making.
- Allow sufficient time for iteration of content development, clearly defining the beneficiaries of the content, and providing clear guidelines on roles, responsibilities, expectations, processes and criteria, which should be indicated upfront to content partners.
- Follow a rigorous, quality-driven content process that embeds quality assurance and process checks within organizational ways of working.
- Allow local content providers to take ownership of content quality to ensure 'buy in' to the benefits of high-quality content and the reputation which supports that.

13.1.3 Involving users

- End-user testing should be established as a crucial component of the content development process.
- The conduct of end-user testing and pre-release of content is essential so that feedback can be incorporated before content is rolled out through new services.

- Creating content based specifically on user-demand is a costly and time-consuming endeavour and requires field research into specific local practices.
- It is essential that sufficient time is allocated to user involvement – including incorporation of learning for users – as part of the content development process.
- Realize that farmers may engage differently when technology is used and to what extent gender, age, literacy and the novelty of the technology influence engagement with the content materials.
- Early prototypes often fail, so use iterative planning cycles and refine ideas based upon feedback from real users.

13.1.4 Incorporation of local languages

- The ability to deliver and exchange content using local languages should be a key requirement for any digital intervention in agricultural settings.
- Where content is written in a local language the additional costs involved in translation to English should be looked at critically and not done purely for quality control or monitoring purposes, particularly when recommended edits involve nuances of language, and corrections to the local language becomes complex and expensive to undertake.
- Where possible, efforts should be diverted from creating 'base' material in English – if this can be sourced or repurposed from elsewhere – thus placing more emphasis on measures to localize.

13.1.5 Targeted information and advice

- Targeting specific information content is better than providing general advice – 'narrowcasting' in preference to 'broadcasting' – delivering targeted information serves farmers' needs better and it is more likely that the farmers will act upon the advice.
- Successful use of digital technologies in agriculture (e.g., Farmerline, Farmforce) is built upon a deep connection to the

problems farmers aim to solve and the commitment to innovate in a unique way that is contextualized to local requirements.

- There are significant constraints on women producers in accessing agricultural resources (e.g., land, credit, technology and markets). Interventions should take account of gendered social relations (including relative power to take decisions and act) that underlie many small-scale agricultural production systems.

13.1.6 Complementary technologies

- Use of digital technologies should complement, rather than substitute for existing channels and sources of information. Avoid pinpointing 'mobile solutions' only and don't neglect the broader information ecology.
- Understand the interplay between different technologies and with non-technology information sources and channels, to develop successful and contextually relevant services.
- A single technology focus causes lack of emphasis on service integration, identifying where mobile phones can be used effectively as part of a mix of technologies.

13.1.7 Behaviour change

- Changes in human behaviour (and therefore overall development) will take longer (years rather than months) than most project development lifecycles, especially if it requires users to re-think the way they conduct day-to-day activities or if it alters existing modes of operation.
- Consider how the transparency effect introduced by mobile technology influences the culture of management of those technologies and systems?
- Always consider the novelty of the content, information, interaction, practices or exchange that is being introduced.
- Offset novelty by relying on trusted intermediaries and popular communication formats (e.g., oral); and enabling people to ask

questions (i.e., have two-way communication). Such formats will be more likely to translate new knowledge and practices into accepted behaviour changes.

13.1.8 Overcoming weak infrastructure

- Weak and unreliable connectivity is likely to be a significant constraint on the implementation of digital technologies in agriculture. This highlights the need for applications to retain full functionality offline, such as for monitoring and data collection.
- Ensure that data recording processes are offline, with the data stored in the device, and then be able to effectively synchronize with permanent data stores when connectivity is available, as well as make use of alternative storage such as USB sticks or via Bluetooth.
- It is critical to understand the environment you are designing for. Does the technology need to be resilient to protect against heat, dust or humidity? Is there a ready power supply, and is it stable? Is the technology affordable, and can it be locally maintained?

13.1.9 Financial sustainability

- Over the long term digital technologies used in agriculture need to be commercially viable, which requires both means of revenue generation and/or alternative funding models.
- Access to finance is a key concern for farmers and the burgeoning availability of mobile-finance should looked at alongside requirements for information services for the agricultural sector.
- Digital technologies can assist proper financial record keeping which can make farmers more creditworthy in the eyes of financial lenders.
- Financial costs for most farmers are prohibitive (such as for those participating in pilot projects) for users of new services which demands low-cost solutions.

- The costs of hardware, software and networking are outweighed by 'non-technology costs' that relate to human resources, training, monitoring and evaluation, as well as the recurrent costs of running innovative digital applications and services.
- The availability of basic, and even feature mobile phones, is not linked with any demographic variable, but income is clearly the variable that strongly influences the adoption of 'smart-phones' or 'tablets', which are required to reap the full benefit of many digital applications and services.
- Many initiatives benefit from initial and ongoing investment in digital infrastructure from donors, and as such, innovation expenses do not constitute fixed costs, they tend to increase over time, as applications need further development and eventual commercialization to make them sustainable.

13.1.10 Dis-intermediation and re-intermediation

- Agricultural intermediaries are found to be crucial for effective implementation.
- The introduction of digital technologies into agricultural sectors can lead to dis-intermediation whereby the position of existing actors (buyers and sellers) in the value chain are threatened.
- Digital technologies may also enable 're-intermediation' – offering new networks of agronomists as an alternative source of support to traditional extension services (more often than not offered by government).
- Farmers are also keen to gain access to digital applications without the need for the intermediaries such as extension workers. The advantages and disadvantages of farmers accessing the apps without intermediaries needs consideration.
- When information is sourced from applications, this may change decision-making processes by farmers, particularly if they have access to the application on their own mobile phone.

13.1.11 Collaboration and partnership

- Interventions that involve digital technologies always involve complex stakeholder arrangements and partnerships. This is a recurring theme in the chapters where open communication (from project design through to implementation) needs to be clear and consistent to achieve effective delivery.
- Facilitate and nurture good relationships with key in-country partners, such as government, service providers and MNOs.
- Communication is often taken for granted and due consideration is required during the design and planning stages to lay out a communication framework, which itself is well communicated and understood by all.
- It is imperative that key stakeholders are engaged in the planning stages in order that expectations of all parties are understood from the outset.
- In line with best practice for project management, flexibility should be built in to cope with changing priorities, whilst adhering to the principles, objectives and scope of existing agreements, and avoiding expensive and time-consuming changes to existing agreements between stakeholders.

13.1.12 Realizing productivity gains

- Agricultural productivity has a broader definition that just crop yield (crop per unit area of cultivated land, and the rate of seed generation). Rather, it should be understood as 'total factor productivity' (TFP) and its two constituents: 'efficiency', arising from reallocation of inputs; and 'technological advancement', arising from changes that are not due to change in the amount of inputs.
- Digital technologies can support productivity for a broad range of factors: agricultural R&D, human capital, infrastructure, institutional development, organizational and process change.
- Digital technologies can further positively impact factors, such as improved agricultural management, compliance,

traceability and more effective use of field officers' time.

- It is also pertinent to ask what happens to the savings from productivity gains or reduced transaction costs. Are more farmers involved in out grower schemes or do the savings translate into higher profits for intermediaries or even into lower export prices that benefit consumers in the Global North rather than producers in the Global South?

13.2 Future Research Questions

The following future research questions are derived from the one-day workshop on *'Digital Technology for Agricultural and Rural Development in the Global South'* that took place on 20 October 2016 at the Global Development Institute (GDI). They are categorized into the key themes that emerged from the workshop and which formed the basis for the structuring of the book chapters. The questions may be of interest to either practitioner researchers in the field or academic researchers in the course of their studies.

Data/information – collection/ storage/conversion

- What happens to all the data collected from farmers?
- What are the ways in which agricultural information and knowledge gets turned into practice?
- How can mobile technologies improve the agency and capabilities of end users to act on content provided through digital channels?
- How can we use big data for agriculture?
- What are the ethical and moral implications of collecting all this data?
- What data protection is there for farmers?
- What are the ethics of sharing personal data?

Technologies/social media

- Which media can maximize the impact of mobiles?
- Do we need to consider social media?
- How can we help farmers to analyse data from social media?

- What are our target users using their smart phones to do?

Behaviour/practice change

- Does information change people's behaviour sufficiently?
- What does existing ethnography teach us about peoples' acceptance of change?
- What should be the design of the interface when the literacy rate is very low?
- How does applying technology change behaviour?
- Are people inventing different practice for themselves?
- Does new practice emerge rather than being planned?
- When no best practice exists – what do you do?

Intermediation/social networks

- What is the risk of technology undermining existing social networks or participatory practices?
- How can these risks be avoided?
- How can ICT enable farmer-to-farmer communication at low cost?
- How do we enable intermediated relationships as opposed to attempting disintermediation?
- What are the knowledge roles of intermediaries in agricultural value chains and how can mobile technologies supplement/ replace these roles?

Institutional/structural change

- What kinds of institutional impacts are produced by content diffusion initiatives?
- What are the relative advantages/disadvantages of disrupting versus sustaining existing structures?
- How can the capacity of state/agricultural infrastructure be strengthened with ICT skills?
- How can 'local intelligence' be channelled into the planning apparatus?
- How does the availability of proper record keeping and real-time agriculture information affect productivity and compliance with standards?

Business/development interface

- What cases/examples/opportunities are there for successfully bridging business and development interests through mobile/ICTs?
- Profit versus Social: which route can provide mobile services at a cheaper cost?
- Where is the IP patent potential? What are the commercial exploitation issues?
- How is the profit motive of mobile service providers hindering development? For example, effect of high transaction costs on rural poor.
- Would using the methods employed in the west – starting bootcamps/launchpads – be better than charitable effort directed development projects?
- Is technology transfer possible in the agricultural sector?
- How does the transparency effect introduced by mobile technology influence management culture and staff qualifications in aggregator companies?
- How can agricultural and rural entrepreneurship initiatives be implemented to attract and benefit the farmers in Africa?

Development outcomes

- Is the use of mobile phones in agriculture really leading to poverty reduction?
- How can people's wellbeing be transformed through agriculture?
- What are the negative impacts of the diffusion of mobiles in the rural context?
- Does mobile lead to real development or are small holders still poor but with mobiles?
- What happens to the savings from reduced transaction cost?
- Are more farmers involved in outgrower schemes or do the savings translate into higher profits for the companies or even into lower export prices that benefit consumers in the Global North?

Reference

Poulton, C., Dorward, A. and Kydd, J. (2010) The future of small farms: new directions for services, institutions and intermediation. *World Development*, 38(10), 1413–1428.

Selected Sources of Further Information and Recommended Reading

Reports

World Bank (2017) *ICT in Agriculture: Connecting Smallholders to Knowledge, Networks, and Institutions.* World Bank, Washington, DC. Available at: https://openknowledge.worldbank.org/handle/10986/27526

CGIAR (2017) Big data for climate-smart agriculture. Climate Change, Agriculture and Food Security (CCAFS). Available at: https://ccafs.cgiar.org/bigdata#.VuNFjfkrLIU

Caine, A., Dorward, P., Clarkson, G., Evans, N., Canales, C. and Stern, D. (2015) Review of mobile applications that involve the use of weather and climate information: their use and potential for smallholder farmers, *CCAFS Working Paper no. 150.* Copenhagen, Denmark: CGIAR Research Program on Climate Change, Agriculture and Food Security (CCAFS). Available at: https://ccafs.cgiar.org/publications/mobile-applications-weather-and-climate-information-their-use-and-potential-smallholder#.Wdd2RrpFyUk (accessed September 2017) or http://hdl.handle.net/10568/69496

FAO (2014) Communication for Rural Development: Guidelines for Planning and Project Formulation. FAO, Rome. Available at: http://www.fao.org/3/a-i4222e.pdf (accessed September 2017).

Vodafone (2011) Connected Agriculture: the role of mobile in driving efficiency and sustainability in the food and agriculture value chain. Accenture/Vodafone, London. Available at: https://www.accenture.com/mu-en/_acnmedia/Accenture/next-gen/reassembling-industry/pdf/Accenture-Connected-Agriculture.pdf (accessed September 2017).

Books

IFC (2013) *Working with Smallholders: a Handbook for Firms Building Sustainable Supply Chains.* IFC Sustainable Business Advisory Service, Washington, DC.

Wenger, E., White, N. and Smith, J. D. (2009) *Digital Habitats: Stewarding Technology for Communities.* CPsquare, Portland, Oregon, USA.

Deneulin, S. (2006) *The Capability Approach and the Praxis of Development.* Palgrave Macmillan, London.

Dorward, P., Clarkson, G. and Stern, R. (2015) *Participatory Integrated Climate Services for Agriculture (PICSA) Field Manual: A step-by-step guide to using PICSA with farmers.* Walker Institute, University of Reading, UK. PICSA manual livestream available at: https://ccafs.cgiar.org/online-launch-participatory-climate-information-services-agriculture-manual#.WWYQw4Tyupo.

Papers/articles

Aker, J.C., Ghosh, I. and Burrell, J. (2016) The promise (and pitfalls) of ICT for agriculture initiatives. *Agricultural Economics* 47, 35–48.

Burrell, J. and Oreglia, E. (2015) The myth of market price information: mobile phones and the application of economic knowledge in ICTD. *Economy and Society* 44, 271–292.

Bell, M. (2015) ICT – powering behaviour change for a brighter agricultural future, *MEAS Discussion Paper*, October 2015, University of Davis, California. Available at: https://agrilinks.org/sites/default/files/resource/files/Bell%20%282015%29%20ICT%20for%20Brighter%20Ag%20Future%20%28MEAS%20Discussion%20Paper%29.pdf (accessed September 2017).

Deichmann, U., Goyal, A. and Mishra, D. (2016) Will digital technologies transform agriculture in developing countries? *Agricultural Economics* 47(S1), 21–33. Available at: http://dx.doi.org/10.1111/agec.12300 (accessed 1 March 2018).

Duncombe, R. (2016) Mobile phones for agricultural and rural development: a literature review and suggestions for future research. *The European Journal of Development Research* 28, 213–235.

Islam, M.S. and Grönlund, Å. (2012) Factors influencing the adoption of mobile phones among farmers in Bangladesh: theories and practices. *International Journal on Advances in ICT for Emerging Regions (ICTer)*, 4(1), 4–14. Available at: https://icter.sljol.info/articles/abstract/10.4038/icter.v4i1.4670/ (accessed 1 March 2018).

Protopop, L. and Shanoyan, A. (2016) Big data and smallholder farmers: big data applications in the agri-food supply chain in developing countries. *International Food and Agribusiness Management Review (Special Issue)*, 19(A), 173–190.

Waugaman, A. (2016) From principle to practice: implementing the principles for digital development. The Principles for Digital development Working Group, Washington, DC, USA. January 2016. Available at: http://digitalprinciples.org/wp-content/uploads/2016/02/mSTAR-Principles_Report-v6.pdf (accessed September 2017).

Websites

Sweet Potato Knowledge Portal available at: http://www.sweetpotatoknowledge.org/ (accessed 1 March 2018).

Index

Page numbers in **bold** type refer to figures and tables.